The Túngara Frog

The Túngara Frog

A Study in Sexual Selection and Communication

Michael J. Ryan

With a Foreword by
Peter Marler

The University of Chicago Press
Chicago and London

MICHAEL J. RYAN is assistant professor of zoology at the University of Texas, Austin, and research associate at the Smithsonian Tropical Research Institute.

The University of Chicago Press, Chicago 60637
The University of Chicago Press, Ltd., London

© 1985 by The University of Chicago
All rights reserved. Published 1985
Printed in the United States of America

94 93 92 91 90 89 88 87 86 85   5 4 3 2 1

Library of Congress Cataloging in Publication Data

Ryan, Michael J. (Michael Joseph), 1953–
  The túngara frog.

  Bibliography: p.
  Includes index.
  1. Physalaemus pustulosus—Behavior. 2. Physalaemus pustulosus—Reproduction. 3. Amphibians—Behavior.
4. Amphibians—Reproduction. 5. Amphibians—Latin America. I. Title.
QL688.E227R93 1985   597.8'7   84–24110
ISBN 0–226–73228–2
ISBN 0–226–73229–0 (pbk.)

To my parents

Contents

	Foreword by Peter Marler	ix
	Preface	xiii
1.	Sexual Selection and Anuran Reproductive Behavior	1
2.	The Species: *Physalaemus pustulosus*	24
3.	Reproductive Behavior	32
4.	Male Reproductive Success	49
5.	The Function of the Advertisement Call	66
6.	Why Do Females Choose Mates?	123
7.	Costs of Reproduction: Energy	143
8.	Costs of Reproduction: Predation	163
9.	Conclusions, Speculations, and Suggestions	186
	Appendix: Methods	191
	References	205
	Index	223

Foreword

This book is about the calling of frogs, and how one can set about analyzing its biological significance. Although Michael Ryan concentrates on one species, his findings have wider relevance, not only for understanding the intricacies of mating behavior in other amphibians, but also for gaining an appreciation of the kinds of evolutionary dynamics to which all sexually reproducing organisms are potentially subject if they must go through the process of attracting a mate. Mate choice is always a crucially important event, and we are provided here with what is probably the most penetrating and comprehensive account ever of its many attendant contingencies and conflicts. Ryan's study goes beyond the basic problem of finding a member of one's own species for a mate, and grapples with more difficult and challenging questions about the extent to which behavior varies between members of the same species, and even within the same population.

This variation, which has rarely been studied, is the raw material upon which the forces of sexual selection can operate. *The Túngara Frog* thus represents a style of ethological research that would have been unthinkable a generation ago, when I was a student. Of the many insights to which the classical ethology of the 1930s gave rise, perhaps most enduring was the simple realization that the natural behavior of animals is, after all, amenable to scientific investigation. It is not necessary to bring animals into highly simplified laboratory situations as a prerequisite for study that is rigorous and quantitative. Konrad Lorenz offered compelling evidence that, far from being infinitely variable, the actions of animals are patterned in clearly defined ways, not just in locomotion, but also in feeding, nest construction, social interactions, and above all in behavior such as the calling of frogs, which figures in social communication.

We are still in debt to our ethological progenitors for the insight that what appears to the uninitiated observer as a series of continuously varying movements, too chaotic to be scientifically manageable, typically proves to have at its core actions that are stereotyped and species-specific. Much early effort was devoted to showing that the structure of so-called fixed action patterns is sufficiently predictable that it can be used as a basis for interpreting the history and relationships of a species, in much the same way a comparative anatomist or taxonomist draws inferences from species differences in the structure of pelvic bones or parts of a jaw. Innate species differences in patterns of action in turn offer the neuroethologist priceless opportunities to develop new insights into how nervous systems maintain control of the fine structure of behavior.

Concepts of classical ethology were actually intended to accommodate notions of both stereotypy and variability, but the focus on species-specificity nevertheless dominated ethological research for thirty years or more. Even when behavioral variations were documented, this was usually more for the sake of rigor and completeness than for testing particular hypotheses about the function and evolutionary significance of behavioral variation. Modern ethology has seen a radical shift in this focus.

In order to progress from simple assertions about the "adaptiveness" of behavior to quantitative demonstrations of the functional significance of variations in the behavior of individuals and populations, precise measurements of patterns of action became not a matter of personal research style but an absolute necessity. The development of genetic theories of social evolution on the one hand, and the application of sophisticated economic theory to the functional analysis of behavior on the other, both conspired to provide new, heuristically fertile approaches that have revolutionized the evolutionary analysis of behavior.

For a time it seemed as though new schisms were developing, severing the old ethology from the new, with the risk of sacrificing some of the progress that had already been made. It is now clear that such fears were unfounded. *The Túngara Frog* serves as an ideal illustration of the scientific advances that can be achieved by blending the best of both, in a broad, multidisciplinary approach.

If one is accustomed to thinking of organisms and their characteristics in typological terms rather than in the manner of a developmental biologist, concepts of innateness, which were intimately linked to those of the fixed-action pattern, seem inevitably to lead to mechanistic interpretations of behavior. Many zoologists took early ethological studies as implying that all animals, except perhaps some higher mammals, can be appropriately thought of as automata, at least most of the time. Amphibians, along with birds, often suffer from this misinterpretation. Thus it will come as a revelation to some readers to discover how complex and variable the behavior of the túngara frog can be. Its song is marvelously intricate, with two distinct parts, each having a distinct function. Ryan is able to explain the interweaving of these functions in beautiful detail. A male frog is in fact able to adjust call complexity according to its circumstances, thus achieving a compromise between maximizing mate attraction and minimizing the risk of predation.

A male frog who is calling by himself gives just "whines." "Chucks" are added to the whines as the level of social activity rises and other males begin to call. A combination of chucks and whines is more attractive to females than the whine alone. Females also tend to favor chucks that are low pitched, and Ryan presents compelling evidence that features of this call have, in fact, evolved in response to sexual selection imposed by the choices that females make between individual males.

If chucks are so attractive to females, why are they not used all the time, whenever a male gives his advertisement call? The answer is that the major predator on the túngara frog is the frog-eating bat. A hungry bat finds that chucks provide ideal cues for locating the frogs on which it preys. That is why, except when calling groups form and social excitement runs high, chucks are withheld, especially if there are any signs of bats in the vicinity. In a group there is also safety in numbers, an individual being relatively less vulnerable in a larger chorus than when alone.

The Túngara Frog provides one of the first quantitatively documented illustrations of the lengths to which an animal will go in balancing the costs and benefits of a particular pattern of behavior, actually adjusting the form of the behavior in dynamic fashion to the circumstance in which it finds itself with remarkable precision. Given the vicissitudes of biological variation, individuals inevitably differ in the success with which they can strike the compromise that offers the best prospects for success, balancing the elicitation of sexual favors against the avoidance of predation, economizing on energy expenditure while maximizing calling time at the breeding site, and constantly modifying the subtleties of their behavior in the process.

These studies give us a new appreciation of the significance of behavioral variability. A consequence of this is the opening up of yet another set of questions for the neuroethologist. Frogs and toads are already favored subjects, having yielded some of the most remarkable insights into the intricate relationships between brain and behavior, which Ryan puts to good use in interpreting the responsiveness of females to low-frequency calls. The highest compliment I can pay to *The Túngara Frog* is that it seems to confirm that neurophysiologists *have* been too inclined to treat amphibians as automata. If, in addition to representing one of the most thorough analyses ever conducted of sexual selection in operation, Ryan's work also encourages neuroethologists to renew the search for plasticity in the frog nervous system, this will help to tie the old and the new ethology together even more securely. I personally would find this satisfying but, more to the point, it would also make for good science.

Peter Marler

Preface

Bizarre structures of animals have captured the attention of naturalists for centuries. Many of these structures share two characteristics: they are used by males to court females, and they reduce the survival ability of the bearer. Not until Darwin proposed his theory of sexual selection was there an adequate understanding of how these structures might evolve.

The theory of sexual selection received relatively little attention during the latter part of the nineteenth century and for most of the twentieth century. But a renewed interest in this theory has arisen. A number of wide-ranging implications of sexual selection are being considered, which has resulted in some rather novel theories. Empirical studies have investigated the ecological correlates of mating systems, the magnitude of the variance of male reproductive success, and the roles of male competition and female choice in determining this variance. But few studies have investigated how individual variation in the sexual displays of males influences variation in mating success among males. This is surprising, since it was the incredible diversity of structures associated with sexual displays that led Darwin to consider the theory of sexual selection.

In this book, I report results of my study of sexual selection and communication in a Neotropical frog—the túngara frog. The central question of this research was simple: Does variation in sexual displays among males (primarily the advertisement call) influence male mating success? Although a number of studies have investigated the role of frog calls in species recognition and the underlying neurophysiological basis for this recognition, individual variation in calls is often considered to be of little biological significance.

To examine how frog calls might have evolved under the influence of sexual selection, it was necessary to document how variation in the call influences male mating success. However, it was also necessary to consider the possible benefits obtained by females that exercise mate choice, the morphological factors that constrain the evolution of call characteristics, how predation risk and energy expenditures influence calling behavior, how the frogs perceive the information in the calls, and the phylogenetic history of the species. Already the number of questions has drastically increased! In this study I attempted to explore as many of the factors that influence the evolution of the sexual display as possible. I was fortunate to be able to collaborate with several outstanding colleagues, and as a result this study provides information from the behavior, ecology, phylogeny, reproductive physiology, and sensory physiology of the species. In this book, I hope that the integration of these data presents a more complete picture of sexual selection and communication than is possible with more restricted avenues of research.

The field of behavioral ecology has gone through a rapid phase of growth, and, as typically occurs in rapidly growing areas of science, theory has outstripped data. In many cases, a phenomenon can be explained by more than one hypothesis. In this book I consider the major hypotheses, derived from theories based on selection and constraints, that have been proposed to explain certain observations. Sometimes the data in hand allow me to reject conclusively all but one of the alternative explanations. In other cases, more than one hypothesis cannot be rejected, but the data suggest that one explanation is more likely to be true. However, often the data do not allow discrimination among several likely alternatives. Then it becomes necessary to consider these more ambiguous results, as well as the more conclusive ones, in order to indicate to the readers the possible alternative explanations, as well as to highlight those data still needed to test theories of sexual selection. This exercise is not very satisfying, but it does reflect the state of the field.

Because explanations of some of the methods employed are rather tedious and are not necessary for an understanding of the results, only the general methods are outlined prior to the report of the results, and the more technical details are given in the appendix.

Since beginning this study in 1978, I have become indebted to a number of people and institutions. I should like to thank some of them, but it is hardly possible to thank all of them. All of the field work reported in this book was conducted in Panama, and most of it on Barro Colorado Island, a field station of the Smithsonian Tropical Research Institute. I am especially grateful to STRI for all of the logistical support, loans of equipment, computer access, and other innumerable favors during my three-year residence on BCI. The Smithsonian Institution provided financial support in the form of a Graduate Student Research Appointment and a Predoctoral Fellowship. I am also grateful to the National Science Foundation (DEB 79-0893), Sigma Xi, and the Gaige Fund for supporting my work on sexual selection in túngara frogs, and to the National Geographic Society for supporting the studies by Merlin Tuttle and myself on bat-frog interactions. The last stages of preparation of the manuscript were completed while I was a fellow of the Miller Institute for Basic Research in Science, at the Museum of Vertebrate Zoology, University of California, Berkely. I am grateful to the Miller Institute and the MVZ for their support.

During the field work, data analysis, and manuscript preparation I have had the privilege of being in residence at three outstanding and stimulating institutions: the Smithsonian Tropical Research Institute, Cornell University, and the University of California, Berkeley. I have benefited greatly, in very different ways, from interactions with colleagues and visitors of all these institutions. In general, I would like to thank the residents and visitors of BCI, and the members of the Behavior Lunch Group at Cornell for their constructive criticisms and patient endurance during the ontogeny of this study. Specifically, I would like to thank Kraig Adler, George Bartholomew, Eliot Brenowitz, Terri Bucher, Bob Capranica, Steve Emlen, Carl Gerhardt, Harry Green, Mark Kirkpatrick, John Phillips, Gordon Rodda, Karen Sherman, Bob Silberglied, Neal Smith, Cindy Taft, Ted Taigen, Pepper Trail, Merlin Tuttle, Dave Wake, Marvalee Wake, Bruce Waldman, Kent Wells, and Mary Jane West Eberhard. All of these people discussed some aspect of this study with me; others also loaned equipment or read portions of the manuscript.

While conducting this study I came to know two very special people, and I would like to acknowledge my sincere thanks and appreciation to them. Stanley Rand began studying túngara frogs in the 1960s and his work provided an important starting point for my study. Stan was always very generous with his time, his ideas, his equipment, and his immense knowledge of tropical biology. He read the entire manuscript and we discussed most of the ideas presented here. I shared many stimulating times, and more than a few bottles of rum with Stan—I thank him for those times and for his friendship. Katherine Troyer provided field and editorial assistance, companionship, and emotional support during much of this study. She read the entire manuscript several times, and the manuscript would not have been completed on time without her assistance. Kathy has been an invaluable friend and I am very grateful for that.

1 Sexual Selection and Anuran Reproductive Behavior

Among many animals the sexes are strikingly different, a condition known as sexual dimorphism. In such species the male is often the more adorned. For example, males of the extinct Irish elk had antlers so huge, some have argued, that they led to the species' demise (e.g., O'Rouke 1970, but see Gould 1974); and male peacocks have elaborate tail plumes that make flight all but impossible. Sexual dimorphism can be behavioral as well: for example, male birds and frogs are usually the more vociferous of the sexes. In fact, the sexes can be so strikingly different that in over a hundred instances conspecific males and females have been erroneously classified as different species (Mayr 1963). Charles Darwin suggested that many of these sexually dimorphic characteristics had evolved under the influence of sexual selection.

This book is a report of my study of sexual selection and communication in the túngara frog, *Physalaemus pustulosus* (Leptodactylidae). My specific interest is the evolution of female mate choice and male sexual displays. But as I will show, this interest has encompassed several lines of investigation, including a detailed analysis of the male's sexual display, measurements of male and female mating success, an estimation of male and female reproductive energetics, sensory physiology, and a consideration of the role of predation in the evolution of sexual displays. A review of some basic concepts of Darwin's sexual selection theory and a brief discussion of how frogs breed will put the results in perspective.

Darwin's Theory of Sexual Selection by Female Choice

Darwin's theory of sexual selection has received renewed interest, especially since the publication of B. Campbell's *Sexual Selection and the Descent of Man* (1972). Consequently, various aspects of sexual selection theory recently have been reviewed by several authors (e.g., Mayr 1972; Halliday 1978; Otte 1979; West Eberhard 1979; O'Donald 1980a; Thornhill 1980,

Bateson 1983). It is not the purpose of this section to provide an exhaustive review of sexual selection theory. Instead, I hope to provide a concise summary of the theory as proposed by Darwin, especially as it concerns the role of female choice, and to briefly review some later extensions and clarifications of the theory. This should enable those not familiar with the theory of sexual selection to appreciate the questions addressed in this research.

Darwin's theory of evolution by natural selection suggested that "if variations useful to any organic being do occur, assuredly individuals thus characterized will have the best chance of being preserved in the struggle for life; and from the strong principles of inheritance, these will tend to produce offspring similarly characterized" (Darwin 1859, p. 98). Thus Darwin proposed a mechanism for organic adaptation to the environment. But his theory of natural selection did not necessarily explain why different sexes of the same species often were quite distinct morphologically. Darwin granted that in some cases, natural selection might be responsible for sexual dimorphism because males and females might be specialized for utilizing different resources. However, many of the sexually dimorphic traits and behaviors possessed by males, such as elaborate plumage, bright coloration, and conspicuous courtship behaviors, seem actually to reduce an animal's survival ability (see review of sexual dimorphism in Selander 1972). In fact, one of the intriguing outcomes of sexual selection is that, in the context of survival ability, it can lead to the evolution of maladaptive traits.

Darwin thought that in many instances sexual dimorphism could be explained by his theory of sexual selection, a concept he introduced in the *Origin of Species:* "This form of selection depends not on a struggle for existence in relation to other organic beings or the external conditions, but on the struggle between individuals of one sex, generally the males, for the possession of the other sex" (Darwin 1859, p. 69). Sexual selection results from variation in the ability of individuals to acquire mates: in sexual selection the organisms are both the agents and the subjects of selection.

Although sexual selection was only briefly introduced in the *Origin of Species*, this theory was meticulously developed in a later work by Darwin, *The Descent of Man and Selection in Relation to Sex* (1871). Here Darwin points out that the sexes differ, by definition, in primary sexual characteristics, and these characteristics have evolved under the influence of natural selection. In a more subtle distinction, he indicated that secondary sexual characteristics that are necessary for reproduction, such as organs found in some crustaceans that are modified for clasping the females, also have been acquired primarily through natural selection. Sexual selection has acted on many secondary sexual characteristics that are quite unconnected with the reproductive organs such as "greater size, strength and pugnacity of the male, weapons of offence, gaudy colouration, power of song" (Darwin 1871, p. 567). These characteristics have evolved not because they increase an individual's ability to survive (they might actually reduce survival ability), but because they increase its chances of mating.

There are two components to sexual selection:

> in the one it is between individuals of one sex, generally the male, in order to drive away or kill their rivals, the female remaining passive; whilst in the other, the struggle is likewise between the individuals of the same sex, in order to excite or charm those of the opposite sex, generally the females, which no longer remain passive but select more agreeable partners. (Darwin 1871, p. 916)

Thus sexual selection results from competition and choice. Usually the competition is among males, and the choice is exercised by females. However, in the seemingly endless variety of organisms in nature, it is not surprising that we sometimes find female-female competition and male choice (e.g., Jenni 1974).

The operational distinction between male competition and female choice is not always obvious, since they are often complexly interrelated, as Darwin appreciated: "But in very many cases the males which conquer their rivals, do not obtain possession of the females, independently of the choice of the latter" (1871, p. 573). For example, some researchers have suggested that a female might behave in a manner that incites competition between males with the result that the female acquires a competitively superior mate (Davies and Halliday 1977).

It was clear to Darwin, and to most of his contemporaries (Ruse 1979), that many of the special weapons found in the animal kingdom, such as tusks and antlers, are restricted to, or more developed in males and have evolved because they increase a male's ability to compete for mates. This is true whether the competition consists of directly obtaining access to females or whether it enables indirect access to mates, for example by allowing males to control territories on which females reside. But the more intriguing aspect of sexual selection was the role of female choice of mates in the evolution of male ornaments. For example, "The case of the Argus pheasant is eminently interesting, because it affords good evidence that the most refined beauty may serve as a sexual charm, and for no other purpose" (Darwin 1871, p. 731; see Davison 1982 for a description of the sexual display of this species). Thus Darwin felt that many of the elaborate traits of males, especially those traits that were obviously of no help in combat and were a hindrance to survival, developed because of their attraction to females.

Darwin appeared to be convinced of the importance of female choice for several reasons. Although few observations had been made of female reproductive behavior in nature (Poulton 1890, as cited in Mayr 1972), Darwin felt that males were the more eager sex and that females were reluctant to mate with just any available male. In an extensive review of much of the animal kingdom, Darwin (1871) showed that in sexually dimorphic species usually the male possessed the more elaborate traits, and often these traits were not developed until sexual maturity. Also, many of these traits, such as the brightly colored breeding plumage of some birds, were only present during the breeding season, and even if they were present all year, they seemed to be displayed only when males were courting females. These observations from the natural world helped convince Darwin of the importance of female choice, and many of these same arguments were put forth by Weismann (1904) in his vigorous defense of sexual selection.

But another factor contributing to Darwin's view of female choice stems from his use of the natural selection/artificial selection analogy (Ruse 1979). Darwin drew heavily on the experience of plant and animal breeders to substantiate both his theories of natural selection and sexual selection. In England at that time, artificial breeding was quickly becoming an important science, as it became necessary for agriculture to increase the yield of its plants and animals in order to meet the demands of an expanding population (Ruse 1979). Not only were breeders able to increase the yield of their animals through artificial selection, but they also increased the beauty of certain species. Darwin was aware of this, particularly because of his interest in domestic pigeons. Darwin suggested that if females preferred to mate with more attractive males in nature, they would similarly increase the beauty of conspecific males over time.

Although the importance of male-male competition in sexual selection was readily accepted, the role of female choice was almost universally and immediately rejected (Mayr 1972)—in part because in order for female choice to operate, females must be endowed with an "aesthetic sense," a capacity most researchers at the time were not ready to accept. Many thought that Darwin was being too anthropomorphic in suggesting that females select males on criteria similar to those used by artificial breeders.

The most important early critic of female choice was Alfred Wallace. Although he acknowledged that male displays might sexually excite females, Wallace suggested that no evidence had been found to show that females preferred the displays of some males over those of others.

> Anyone who reads these most interesting chapters will admit, that the fact of the display is demonstrated; and it may be also admitted, as highly probable, that the female is pleased or excited by the display. But it by no means follows that slight differences in the shape, pattern, or colour of the ornamental plumes are what lead a female to give preference to one male over another (Wallace 1905, p. 285)

Wallace offered a number of alternative explanations for the evolution of sexually dimorphic characters. For example: female birds are less brilliantly colored than males because they must be camouflaged while incubating the eggs (a theory recently discussed by Baker and Parker 1979); many male traits thought to have evolved under the influence of sexual selection act only as species-recognition mechanisms, but do not influence female mate choice among conspecifics; and some exaggerated male traits might be due to a surplus of vital energy which leads to abnormal growth. With the exception of the last suggestion, Wallace raised some important criticisms of sexual selection theory which highlighted the need for carefully controlled experiments examining how females select mates. However, Ruse (1979) has suggested that the difference of opinion about female choice between Darwin and Wallace might have been primarily because the analogy of female choice was striking in artificial selection, an area often invoked by Darwin to substantiate his theories, but an area not often investigated by Wallace.

Without doubt, the most influential critic of Darwin's theory of sexual selection by female choice was Julian Huxley: "it has now become clear that the hypothesis of female choice and of selection between rival males irrespective of general biological advantage is inapplicable to the great majority of display characters" (Huxley 1938b, pp. 20–21; see also Huxley 1938a). But Huxley's discussion of female choice is sometimes confusing. For example, he reviews a number of studies and observations suggesting that sexual displays may act to synchronize the physiological condition of the sexes. He then concludes, "The facts recorded in the previous section indicate that display may often be of advantage to the species in promoting more efficient reproduction. Any resultant selection will therefore come under the head of Natural Selection, not Sexual Selection in Darwin's sense" (Huxley 1938b, p. 20). By this argument, Huxley interprets male displays as a species adaptation. This implies that natural selection is acting at the level of the group, rather than at the individual level that most biologists assume is the more common occurrence (e.g., Williams 1966). It is not likely that male sexual displays are a result of group selection.

Another problem is Huxley's dismissal of female choice as an agent of sexual selection merely because female choice has a proximate cause; he suggests that it is a physiological response to the male's display. But if the displays of some males are better able to elicit this physiological response from females, then female choice could influence the evolution of the male display in accordance with Darwin's hypothesis. This argument certainly does not demonstrate the efficacy of female choice, but it suggests that Huxley dismissed the possibility for fallacious reasons. (O'Donald 1980a offers an extensive critique of Huxley's views of female choice.)

Huxley and Wallace were not alone in their concern for the lack of evidence demonstrating that male sexual displays are important in female mate choice among conspecifics. Noble and Bradley surveyed the reproductive behavior of lizards from seven families in an effort to determine whether coloration and displays of males functioned in male-male interactions or in attracting females. Their conclusions were unqualified:

> The bright colors which adorn males of many lizards do not serve as attracting devices as has been assumed hitherto. Neither females

nor males are attracted by the displays. Bright colors have the important function of aiding in sham fights when the males attempt to avoid combat with rivals by making themselves appear as large and conspicuous as possible. The greatest displays of colors are directed toward rival males, and not toward females. (Noble and Bradley 1933, p. 94)

Interestingly, although their results directly challenged the hypothesis of female choice, Darwin's work is never cited in the paper by Noble and Bradley.

The important criticism of the theory of sexual selection by female choice made by Wallace, Huxley, as well as many of their contemporaries, were based on two main objections: (1) little evidence exists to show that the sexually dimorphic characters of males are used to attract females, as opposed to functioning in male combat; (2) even if one accepts the notion that male traits do function in eliciting sexual behavior from females, little evidence exists to show that variation in the male traits leads to variation in their ability to acquire mates. Some of the other criticisms of female choice were either trivial or wrong. But these two points are extremely important. So, some sixty years after Darwin presented his sexual selection theory in full form, the importance of female choice in the evolution of male sexual displays was not widely accepted.

Fisher's Contribution to Sexual Selection Theory and Recent Extensions of the Theory

Ronald Fisher made the most important post-Darwinian contribution to sexual selection theory in 1930, when he elaborated Darwin's ideas on female choice and developed the theory of runaway sexual selection. Like Darwin, Fisher realized that the evolution of male ornaments and female preference would proceed at an accelerating rate due to the increased mating success experienced by sons of "choosy" females. These male offspring would inherit the characteristics that made their fathers attractive and therefore would also be more likely to attract females. Male ornaments and female choice thus coevolve through a type of "self-reinforcing choice" (Maynard Smith 1978), and continue to do so until opposed by the forces of natural selection. Darwin and Fisher both emphasized this accelerating nature of sexual selection:

> In regard to structures acquired through ordinary or natural selection, there is in most cases, as long as the conditions of life remain the same, a limit to the amount of advantageous modification in relation to special purposes; but in regard to structures adapted to make one male victorious over another, either in fighting or in charming the female, there is no definite limit to the amount of advantageous modification. (Darwin 1871, p. 583)

This distinction between natural and sexual selection is an important concept, perhaps not appreciated by some authors (e.g., Otte 1979) who have suggested that the distinction between natural and sexual selection is not warranted. (See West Eberhard 1979 for a discussion of the relationship between sexual selection and natural selection.)

However, Darwin did not clearly explain why some males should be more attractive to females than other males, and especially why over evolutionary time females should continue to prefer males that are maladaptive in the struggle for survival. Surely the initial females would not receive the benefit of having attractive sons if the preference for those male characteristics were not a common female behavior. Fisher solidified Darwin's theory by suggesting that in the initial stages of the coevolution of male traits and female preference, females must gain some benefit other than having more attractive sons. For example, females might initially select males with more plumage because it demonstrates that these males have some advantage in terms of natural selection. Perhaps they are more healthy, vigorous, or able to acquire food. If a genetic basis to these traits exists, then females choosing males will benefit because they will have healthy sons, and the daughters will inherit the preference for these males. Once this preference for more plumage has been established in the population, the sons of choosy females would benefit because they would inherit those traits that females preferentially choose. Sexual selection is off and running!

Lande (1980, 1981) and Kirkpatrick (1982) have presented a genetic model of Fisher's runaway sexual selection theory, and their treatments have resulted in several nonobvious interpretations. If we assume that there is heritable genetic variation of female preference and male traits, as has been argued by Lande (1976, 1980), then when a female preference initially arises it will result in a genetic covariance between the alleles that determine the female preference and those that determine the male trait. For simplicity's sake, consider two sets of alleles in a species with only one set of chromosomes. L and L' influence tail length in males with L' resulting in longer tails. P and P' influence female mating preference. P females mate at random, but P' females prefer to mate with males having longer tails. The offspring of P' females will have both the P' and L' alleles, thus exhibiting longer tails if they are males and the preference for longer tails if they are females. Because P' females mate preferentially with L' males, this combination of alleles will occur in offspring more frequently than if mating were random. This condition is known as linkage disequilibrium, which results in coevolution of the female preference and the male trait, as suggested by Fisher.

These models show that when the system equilibrates, a given level of female preference will result in a given male phenotype. If the optimum phenotype under sexual selection and natural selection are not identical, then the resulting male phenotype will be a compromise between these two forces. This equilibrium is unstable, in the sense that it is not unique but depends on the starting frequencies of the male phenotype and the female preference. An important conclusion of these genetic models, and one that corroborates the crucial factor of Fisher's theory, is that the female preference maintains the male phenotype away from the natural selection optimum; that is, female choice can result in the evolution of maladaptive (vis à vis natural selection) traits.

These recent models of sexual selection differ with Fisher's model on several points. Lande and Kirkpatrick suggest that an initial natural selection advantage to females exhibiting the preference is not necessary to initiate the runaway process. Once linkage disequilibrium occurs, for whatever reasons, the runaway process will begin. This is also true if the male trait and the female preference are determined by the identical alleles. For example, it is thought that in some crickets both the species-specific pattern of the male's song as well as the female's preference for this song pattern might have the same genetic basis (Hoy and Paul 1973). If this is true, then a genetic covariance between the preference and the trait must exist, and both will be subject to runaway selection. Another important result of this genetic covariance between the sexes is that an evolutionary change in the male trait due to forces other than sexual selection, such as natural selection or genetic drift, will necessarily result in a corresponding change in the female preference. However, it should be noted that O'Donald (1983) criticizes some aspects of Lande's assumptions about the constancy of linkage disequilibrium.

Lande and Kirkpatrick differ with previous interpretations of sexual selection by Fisher and others (e.g., O'Donald 1980a) on another crucial point. Fisher suggested that the male trait would continue to evolve under the influence of sexual selection until opposed by the forces of natural selection. But the genetic models suggest that sexual selection can continue to operate until the survival ability of males is decreased to such an extent that the species is driven to extinction. Lande and Kirkpatrick also suggest that the direction of the evolution of the trait is indeterminate. In the absence of other forces, the elaboration of the trait is just as likely to increase as it is to decrease. These models offer some exciting new views on the dynamics of sexual selection and should provide a focal point for future empirical research (see also Arnold 1983).

Although many aspects of sexual selection theory have been both documented and qualified (e.g., Bateman 1948; Maynard Smith 1956; O'Donald 1962, 1967, 1972, 1973, 1978, 1980b; Trivers 1972) the most radical change in the concept since Fisher has been proposed by Zahavi (1975, 1977). Briefly, Zahavi emphasizes that the elaborate morphology and behavior associated with male courtship represent a cost. This cost is reduced survival, due perhaps to increased predation, increased energy expenditures, decreased mobility, etc. The idea is not a new one. As I have pointed out, this obvious cost of male ornaments suggested to Darwin that their evolution by natural selection was impossible. Zahavi suggests that male sexual ornaments, either behavioral or morphological, represent a "handicap," and that females use the handicap to assess male survival ability. Individuals that have survived with a handicap have demonstrated their superiority. Males without handicaps may be of similar quality, but they have not been tested in a similar way, so females cannot judge their quality. Therefore, females should prefer handicapped males as mates. Zahavi differs with Fisher on a critical point, namely, that the handicap principle does not allow for self-reinforcing choice: the offspring of choosy females do not benefit solely because they become more attractive mates.

A basic problem with the handicap principle is that when females choose handicapped males, their sons inherit the father's handicap and, consequently, the sons' survival ability is reduced. Genetic models do not corroborate Zahavi's theory, and the criticisms from some population geneticists have been quite vigorous (e.g., Maynard Smith 1976; Davis and O'Donald 1976; see especially O'Donald 1980a). But others have suggested that there might be some credence to the handicap model, although in a modified form (e.g., West Eberhard 1979; Anderrson 1982b; Dominey 1983). At this time, it appears reasonable to conclude that the handicap principle, in the precise terms proposed by Zahavi, does not seem to present a viable alternative explanation for the evolution of sexually dimorphic traits.

Before we continue this discussion of female choice, it is important to state explicitly some of the genetic assumptions of this theory and to clearly define the difference between selection and an evolutionary response to selection. A trait will only evolve in response to selection if that trait varies among individuals in the population and if there is a genetic component to that variation. The genetic component of the variance of a trait is called heritability. It is calculated by dividing the variance of the trait that is due only to additive (i.e., nondominance or nonepistatic interactions) genetic effects by the total phenotypic variance of the trait. The phenotypic variance includes the variance due to genetic effects plus the variance due to environmental effects (e.g., Lewontin 1970).

When we say that there is sexual selection for a given male trait, we mean that males possessing that trait experience greater than average reproductive success, which says nothing about whether or not the trait will evolve in response to sexual selection. Selection will proceed in the absence of any heritability for the trait being selected. But evolution of the trait can only occur in the presence of heritable genetic variation. This concept is basic to evolutionary theory, and its importance to sexual selection has been emphasized by Wade and Arnold (1980).

An important criticism of the theory of sexual selection by female choice is that there is little or no heritability for male traits under intense sexual selection. In the absence of forces such as mutation, which increase the genetic variability within a population, constant selection for any trait eventually will drastically reduce the heritability of that trait. This fact is well known to artificial breeders attempting to increase the productivity of poultry or dairy cows (Falconer 1981). It has been argued that female choice should have the same effect on the heritability of a male trait. For example, if females always mated with males possessing alleles that resulted in the males having long tails, then eventually these alleles would become fixed in the population, and all males would have long tails. Still some variation in tail length among males would occur, but all of the variation would be due to environmental effects and would have no genetic basis. The heritability of the male trait would be zero. Sexual selection will proceed, however, as females continue to mate with males having longer tails. But no evolutionary change in tail length will occur because mated and unmated males do not differ genetically with respect to tail length. Therefore, even though female choice based on variation in a male phenotypic trait can be demonstrated, evolutionary change in the male trait is not necessarily dictated (e.g., Searcy 1982). Others have suggested that since female preferences and male traits probably are controlled by a number of alleles (i.e., they are polygenic), normal mutation rates will maintain sufficient heritability to result in evolution (e.g., Lande 1976, 1980). This might be the most important question about sexual selection that is currently being debated; we will return to this issue in chapter 6. Suffice it to say here that even when sexual selection is demonstrated, evolution in response to that selection will occur only if there is heritability for the traits being selected.

Criteria of Female Choice

Fisher firmly established the importance of female choice in sexual selection. The volume *Sexual Selection and the Descent of Man*, edited by B. Campbell and published in 1972, also presents strong evidence for the efficacy of female choice (e.g., Ehrman 1972; Mayr 1972, 1982; Selander 1972; Trivers 1972). Although the idea of female choice has become accepted, still how or why females select mates is not always clear. For example, a number of studies and discussions of avian mating systems have indicated that females prefer to mate with males on territories with more or better resources (J. L. Brown 1964; Orians 1969; Pleszczynska 1978; Searcy 1979; Lenington 1980). But determining the extent to which females are choosing the mate or choosing the territory is difficult (Searcy 1979). A similar problem in evaluating female choice exists in studies of some species in which males advertise for mates at communal display areas known as leks. The lek is removed from the feeding and nesting area, so in these mating systems resource quality clearly does not influence a female's choice of mates. Females often seem to prefer males that advertise from certain locations in the leks, sometimes known as mating centers (Hogan-Warburg 1966; Kruijt and Hogan 1967; Buechner and Roth 1974; Bradbury 1977). Therefore, it is not clear whether females base their choice on the male per se, or if instead they mate with any male that happens to be at the mating center (see review by Wrangham 1980).

In terms of the evolution of the male's traits in response to sexual selection, it is not important whether females prefer certain male traits, or if females instead prefer certain territories or mating localities that happen to be controlled by males with certain traits. In either case selection for males with specific traits occurs because these males have increased reproductive success. Given sufficient heritabilities for the traits being selected, the traits should evolve in response to sexual selection. However, an assumption crucial to Darwin's and Fisher's hypotheses about the coevolution of male traits and female choice is that females preferentially select mates based on characteristics of the male itself.

Darwin (1871), Weismann (1904), and Fisher (1958) all realized that direct observations of females selecting males were meager. In fact, in a later defense of sexual selection, Darwin argued that females did not necessarily base their choice on specific male characteristics: "I presume that no supporter of the principle of sexual selection believes that females select particular points of beauty in the males; they are merely excited or attracted in a greater degree by one male than another" (Darwin [1876], 1977, p. 210). And later, in discussing observations of mate choice in street dogs, Darwin was not at all convinced that the precise preference of females could be identified. "What the attractions may be which give an advantage to certain males in wooing . . . can rarely have been conjectured" (Darwin [1882], 1977, p. 279). Darwin's skepticism aside, several authors (e.g., Mayr 1972; Davies 1978; Maynard Smith 1978; Thornhill 1980) have indicated the necessity of determining the precise criteria by which females choose males and have voiced their concern with the lack of data supporting such a central concept of sexual selection theory.

Goals of the Research

This brief review of the theory of sexual selection by female choice should indicate the questions that need to be addressed for a critical evaluation of the theory. The first question is: Do females behave in a way that determines with whom they mate? Because the pairing of the sexes may result from a number of causes, such as male competition or random encounters, it is necessary to make detailed observations of the behavior of both sexes prior to mating. Without these observations it is impossible to delineate the relative importance of female choice as an agent of sexual selection. In chapter 3 I provide observations suggesting the importance of female choice in the mating system of túngara frogs.

By definition, selection only occurs if some subset of the population has greater reproductive success. If female choice results in sexual selection, then females must preferentially select a subpopulation of the males as mates. Therefore, one must quantify male mating success to answer the question, Does female choice influence male mating success? This question is addressed in chapter 4.

Does the male's sexual display have anything to do with female choice among conspecific males? This question echoes the concerns of Wallace, Huxley, Noble, Bradley, and many others. To answer in the affirmative, some variation in male displays must be shown, the males exhibiting a subset of this variation must have greater mating success, and it must be demonstrated, preferably experimentally, that females have the capabilities to discriminate among display variants and that they preferentially select the variant that is correlated with greater male mating success in nature. A rejection of any of these assertions must lead to the rejection of the hypothesis that female choice based on male traits influences male mating success. In chapter 5, I examine how variation in the male sexual display of túngara frogs influences male mating success. This examination necessitates a detailed investigation of the structure and function of the sexual display, which also is considered in chapter 5.

As an evolutionary biologist, I am concerned with the processes that have given rise to reproductive behaviors. In chapter 5 I review phylogenetic evidence suggesting that female choice might have influenced the evolution of the sexual display of túngara frogs, and I consider morphological constraints on the evolution of the sexual display. Chapter 6 addresses three possible explanations for the evolution of the particular type of mate choice exhibited by female túngara frogs. (1) Female mate choice is a consequence of the structure of the auditory system. (2) Female mate choice evolved under the influence of natural selection because it either insures mating with conspecifics (and thus with males that have compatible genomes), increases the reproductive output of the female, or results in females producing offspring that are better adapted for survival. (3) Female mate choice evolved through the process of sexual selection, coevolving with the male trait.

Reproductive behaviors, especially male sexual displays, can only evolve under certain constraints. Two forces that greatly limit the extent to which sexual selection can act are predation and energy expenditure. In chapters 7 and 8 I review data on how predation on calling males, especially predation by bats, and how energy expenditures for reproduction by both sexes influence the reproductive patterns observed in nature.

In summary, the main question addressed in this study are: (1) Does female mate choice influence male mating success? (2) Which males do females select as mates? (3) On what characteristics do females base their choice? (4) What selective advantages, if any, are obtained by females choosing certain males? and (5) How do the associated costs of reproduction, predation, and energy expenditure influence the species' reproductive patterns?

Anuran Breeding Patterns

My study of sexual selection and communication was conducted with frogs, a group of animals that comprise the order Anura. For those not familiar with these marvelous little creatures, I will review some of the basic characteristics of their mating systems.

The reproductive behavior of anurans was once considered in rather simplistic terms. "If it is not small enough to eat nor large enough to eat you, and doesn't put up a squawk about it, mate with it" (Jameson 1955) is a statement that at one time characterized much of the thinking on this subject. But the complexity of anuran social behavior, and especially their mating systems, is now recognized by a number of authors (e.g., see symposium volumes with sections discussing anuran reproductive behavior: Blair 1972; Vial 1973; Taylor and Guttman 1977). The review of anuran social behavior by Wells (1977a) was important in synthesizing the literature at that time in terms of current social-behavior theory.

For most species of frogs, breeding usually takes place in water. Males typically vocalize during the night, and females are attracted to the breeding site by the calling males. Once females are at the breeding site, mating eventually takes place. The duration of the mating act is much longer for frogs than for most other vertebrates, lasting anywhere from hours to weeks or months (Wells 1977a). Prior to fertilization, the male clasps the female from above, head to head with his underside resting against her back. This clasping behavior is referred to as amplexus. Sometime during amplexus, the female begins to extrude eggs from her cloaca. As the eggs pass by, the male releases sperm from his cloaca, fertilizing the eggs. The most interesting aspect of the frog's reproductive behavior takes place after the females arrive at the breeding site and before amplexus occurs. Many studies and theories have examined how pairing occurs.

The basic tenet of Wells's (1977a) review is that most social interactions among anurans occur during the breeding season and that the predominant feature of anuran mating systems is that males vie for mates. He suggests that two basic mate acquisition behaviors occur among frogs: scramble competition and female choice. Scramble competition consists of males moving about at the breeding site and clasping anything that resembles a female. The goal of this behavior appears to be the discovery of a female through trial-and-error clasping. In the early spring, males of some species such as wood frogs can be seen frantically clasping a variety of very inappropriate objects, such as salamanders and beer cans. In contrast, when female mate choice is a predominant feature of the mating system, males usually remain stationary at their calling sites. Females move freely through the breeding site until they initiate mating by making physical contact with a male. Wells proposes that the spatial and temporal distribution of females determines the form of mate competition. This concept also has been applied by Emlen and Oring (1977) in their discussion of avian mating systems.

Anuran species can be characterized by the length of their breeding seasons: species with short breeding seasons (explosive breeders) are at one end of the continuum, and those with long breeding seasons (prolonged breeders) are at the other. In explosive breeders, the female distribution of the population is clumped in both space and time. Males of these species usually acquire mates through scramble competition; they actively search for females. Wells states that in species in which males form dense aggregations, females do not approach calling males because any female attempting to do so would be intercepted by a searching male.

In prolonged breeders, however, gravid females seek out vocalizing males and select them as mates. Wells lists 16 species in which the female approaches or seems to approach the male. As studies of anuran reproductive behavior have intensified, this type of female mate-choice behavior has been reported in a number of additional species. In these species, the vocalization attracts the female not only to the breeding site, but to her future mate as well. In some species with prolonged breeding seasons, there can be "peripheral" males: they do not call but instead adopt a noncalling mating behavior. Wells suggests that in many species these noncalling males probably are not able to intercept females, but remain in the noncalling posture until they can take over a territory or a calling site abandoned by a mated male (see also Fellers 1975). Others have demonstrated that in some species the interception of females en route to calling males is the primary function of the noncalling behavior (Perrill, Gerhardt, and Daniels 1978, 1982).

The dichotomy between mate acquisition behaviors, Wells suggests, arises from the necessity of maximizing either speed or energetic efficiency. In species with very short breeding seasons, females are available for only a short period of time thus there is a need for males to acquire mates rapidly. Males of prolonged breeding species, on the other hand, should be expected to maximize energetic efficiency in order to take advantage of female receptivity which is spread over a long period of time. The more conservatively a male expends energy for reproduction, the more time he can spend at the breeding site. Of course this dichotomy expresses the two possible extremes; a male's mate acquisition behavior should fall within a continuum between scramble competition and female attraction.

These different mate acquisition behaviors should influence the degree to which sexual selection occurs. The intensity of sexual selection is a function of the variance of male reproductive success, which results from the differential ability of males to acquire mates (Wade and Arnold 1980). Since different mating systems are characterized by different variances of male mating success, the importance of sexual selection in the evolution of male mating behavior is expected to vary among mating systems.

Wells suggests that the synchronous breeding exhibited by females with explosive breeding seasons reduces the possibility that a few males will obtain most of the matings. This is especially true for anurans, since in most cases a male can only mate once per night due to the prolonged duration of amplexus. But in species with an extended breeding season, female receptivity occurs over a longer time period, and a greater opportunity exists for a few males to acquire most of the matings. Thus the distribution of male mating success is likely to be more skewed (i.e., has a larger variance) in prolonged breeders than in explosive breeders. Consequently, a greater potential should exist for sexual selection to influence the evolution of male reproductive behaviors in prolonged breeding species than in explosively breeding species.

Advantages of Studying Sexual Selection and Communication with Anurans

Many aspects of their reproductive biology make anurans excellent subjects for investigating questions of sexual selection and communication. Male frogs usually advertise for mates by vocalizing and thereby reveal their presence to the observer. Also, many species form dense aggregations while breeding, enabling the researcher to observe a number of individuals simultaneously. Frogs can be marked permanently by toe clipping, and they usually can be fitted with some type of marker, such as a numbered waistband (Emlen 1968a), that allows undisturbed behavioral observations. Another advantage of studying sexual selection in frogs, and one that cannot be overemphasized, is the ability of the researcher to assign paternity. Fertilization is usually external, and in most species there is no doubt as to which male fertilizes the eggs. So in most cases the researcher can quantify male mating success with complete confidence. This level of confidence does not exist in studies of many other taxa. For example, some studies have estimated a male's reproductive success by the number of times he copulates, but Hausfater (1975) has shown that in baboons the timing of copulation within the female's estrous cycle is a more reliable measure of male fertilization success than is the absolute number of copulations: a male that mates with a female close to her time of ovulation is more likely to achieve fertilization. Other studies have relied on the number of females in a male's territory or in his harem as an indicator of male reproductive success, but Bray, Kennedy, and Guarino (1975) found that females in the territories of vasectomized red-winged blackbird males produced fertile eggs. Similarly, Tannenbaum (1974) demonstrated that when a harem leader of the white-lined bat was vasectomized, some females in his harem still produced offspring.

Anurans are also excellent subjects for investigating the role of the sexual display in mate choice. The primary sexual display—the advertisement call—is conspicuous and can be tape-recorded and analyzed using modern bioacoustic techniques. Therefore, any correlation between male mating success and characteristics of the sexual display can be determined. Because calls can be synthesized and because females readily approach speakers broadcasting the conspecific call, hypotheses about the role of certain sexual display characteristics in mate choice can be tested experimentally. These techniques already have proven successful in investigations of the role of the advertisement call as a species-isolating mechanism (e.g., Blair and Littlejohn 1959; Bogert 1960; Blair 1964; see also the review in Salthe and Mecham 1974) and in studies of the sensory basis of species discrimination (e.g., Gerhardt 1974, 1978a, 1978b, 1981a, 1981b; Loftus-Hills and Littlejohn 1971).

A final advantage in studying the anuran sexual display is the detailed understanding of the anuran auditory system, resulting chiefly from the investigations of Capranica and his colleagues (see reviews by Capranica 1976a, 1976b, 1977). This work has resulted in a number of studies examining the neurobiological basis of anuran sexual displays (e.g., Capranica, Frischkopf, and Nevo 1973; Narins and Capranica 1976, 1978). The results of these studies can serve to define the sensory limits within which we can expect the communication system to operate.

Do Female Frogs Choose Their Mates?

Most toads of the genus *Bufo* are characterized by very short breeding seasons (i.e., explosive breeders). But contrary to the predictions made by Wells (1977a), a number of studies have suggested that variance in male mating success in some toads is explained primarily by female choice of certain sized males, and that females distinguish a male's size on the basis of the frequency of the male's call. Licht (1976, *B. americanus*) and Davies and Halliday (1977, *B. bufo*) suggested that females choose males of a size appropriate to effect a closer juxtaposition of the pair's cloacas during fertilization. Wilbur, Rubenstein, and Fairchild (1978, *B. quercicus*) suggested that females choose the largest males. Is this size bias in male mating success due to females being preferentially attracted to certain types of calls, as these authors suggest? Although a correlation usually exists between male size and call frequency for a variety of frog species (Blair 1964; Zweifel 1968; Martin 1972; Oldham and Gerhardt 1975; Duellman and Pyles 1983; see also chap. 5), Zweifel showed this was not the case in all populations. Of the above studies of *Bufo* mate choice, only Davies and Halliday (1978) showed that male size is correlated with call frequency. However, none of these researchers observed the behavioral interactions that led to mating; amplexed pairs were merely collected at various times after mating was initiated. Therefore, no observations support the conclusion that female choice, as opposed to male-male competition, is primarily responsible for the resultant pairings. This is an important oversight. As indicated above, Wells (1977a) suggests that a major determinant of male mating success in species with short breeding seasons should be scramble competition, and some studies do show that larger male toads are more successful in directly competing for females (Davies and Halliday 1978, 1979). The studies just discussed, however, fail to discriminate between the effects of male-male competition and female choice.

Fairchild (1981) has suggested that female *B. fowleri* are attracted preferentially to calls of larger males. He presented females with calls that were manipulated in such a way so as to correspond to males of different sizes. When given a choice, the females approached the calls corresponding to the larger males. But Fairchild does not offer any behavioral observations indicating that females choose mates; instead he cites Wilbur, Rubenstein, and Fairchild (1978) as demonstrating the importance of female choice in determining male mating success in *B. americanus*. As pointed out above, however, Wilbur, Rubenstein, and Fairchild presented no observations or data germane to this point. Therefore, even if females exhibit preference in the laboratory, it is not clear if they would be able to excercise this preference in the field as well. Furthermore, Fairchild does not even indicate if call parameters vary significantly with male size, and if so what is the nature of the relationship. Fairchild's study does not seem to support his contention that female *B. fowleri* select larger males based on call characteristics (also see Sullivan 1982b; Christian and Tracy 1983).

Clearly, more detailed observations are needed of reproductive behavior in toads with explosive breeding seasons, and these observations now seem to be forthcoming. Sullivan (1983) has presented detailed observations of the reproductive behavior of the explosively breeding toad *Bufo cognatus*. He reported that in the field females are attracted to the vocalizations of males, even though they are sometimes intercepted by noncalling males. The females move among the calling males and initiate amplexus by making physical contact with them. Females also exhibit a variety of behaviors while approaching calling males which results in their avoidance of noncalling males. Almost identical behaviors are reported for female *B. americanus* by Waldman (n.d.), and Gatz (1981) also presented some limited data suggesting that vocalizations by males and mate choice by females are important determinants of male mating success in toads.

These studies suggest that the role of female choice in *Bufo* mating systems should not be dismissed out of hand. Because male-male competition is the most obvious component of the mating system in many *Bufo* and in other explosive breeders (e.g., *Rana sylvatica,* Howard 1980; Berven 1981), the elucidation of the role of female choice may be difficult, but not necessarily impossible.

Several studies of species with prolonged breeding seasons have shown that females select their mates. In these species males vocalize from stationary calling sites, and scramble competition among males, if it exists at all, is not as prevalent in prolonged breeders as it is in explosive breeders.

Females are attracted to calling males and initiate amplexus by making physical contact with the calling male. Females have been observed to bypass some calling males en route to others (e.g., Emlen 1968b, 1976). In some species of frogs with prolonged breeding seasons, males compete for and defend territories that are used as oviposition sites by females. In green frogs (*Rana clamitans*), Wells (1977b) found a strong correlation between his measure of the quality of the male's territory and the male's mating success, and a significant but weaker correlation between male size and mating success. Wells suggested that females preferentially selected males on the basis of territory quality, and that better territories were acquired differentially by larger males. Bullfrog (*Rana catesbeiana*) females also chose larger males on the basis of better territories, and Howard (1978a, 1978b) showed that eggs deposited in territories of larger males had fewer developmental abnormalities and suffered less predation. He also suggested that females that chose larger males gain genetic benefits for their offspring through natural selection. Since larger males are older, they have demonstrated the adaptiveness of their genotypes for survival; these adaptive traits should be passed on to their offspring, although these benefits might be insignificant relative to those provided by higher quality territories (Howard 1978a).

These studies collectively demonstrate the importance of female choice in determining male mating success. In addition, mate choice may also increase a female's reproductive success. These studies were not designed to determine the mechanism of female choice, such as whether choice is based on characteristics of the male or the territory. It has been argued that when territorial quality has an immediate effect on reproductive success, females should be expected to base their choice on the territory rather than on the male, unless higher quality males consistently defend highly quality territories (Searcy 1979). This argument is based on two obvious assumptions: (1) territorial quality is more easily assessed by the female than is male quality; and (2) if females gain any genetic benefit from males, it probably is less than the benefit they would gain by mating with any male in a high quality territory. If these assumptions are correct, the role of male traits in female choice by a territorial species would be difficult to uncover. That male traits are more important in female mate choice seems likely in those species in which males do not control resources of direct benefit to females, such as lekking birds (see also Maynard Smith 1978).

Some studies do suggest that variation in sexual displays of male frogs influence female mate choice. For example, in some species females preferentially select those males that initiate bouts of calling. Whitney and Krebs (1975) noted that in *Hyla regilla*, the frog that initiated calling was also the last individual to stop calling, called at a faster rate, and produced more intense calls. Limited observations in the field showed that females selected bout leaders as mates. Whitney and Krebs tested the influence of initiating bouts on male mating success by allowing a female to choose among four speakers broadcasting calls antiphonally, with one speaker initiating the bout with three calls. Females did preferentially choose bout leaders in these experiments. Whether the same males of *H. regilla* remained bout leaders from night to night is not clear.

Wells (1977a) criticized the conclusion of Whitney and Krebs, suggesting that even if chorus leaders do have higher mating success, this does not necessarily imply that females "prefer" these males because they make "better mates." He states that these males might just be easier to locate in a chorus because they call more. But regardless of the mechanism underlying female response, by mating with males that are easier to locate, females might possibly gain a selective advantage, as suggested by Fisher, because they will produce offspring with these same traits. Since their sons might enjoy greater mating success if the females mate with males that are easier to locate, then in terms of sexual selection, females actually are choosing males that make better mates. This idea, of course, assumes some heritability of the male trait. Although there is a heritable component to the amount of calling in other animals (e.g., crickets; Cade 1981), no data are available that allow us to evaluate this possibility for *H. regilla*. However, this study does clearly demonstrate that variation in male sexual display can lead to variation in male mating success. Similarly, studies of toads have also shown that females are attracted preferentially to males that call more often (Sullivan 1982a; Arak 1983a).

Brattstrom and Yarnell (1968) argued that chorus leaders have greater mating success in *Engystomops* (= *Physalaemus*) *pustulosus*. They observed males calling in captivity and concluded that there was a consistent chorus leader, that the sequence of calling was the result of a dominance hierarchy, and that chorus leaders had greater mating success because they defended territories that were preferred by females as nesting sites. The generality of their conclusions is questioned because in the field, there are no consistent chorus leaders throughout the night, and mated pairs do not deposit nests at the male's calling site (Sexton and Ortleb 1966; chap. 3).

This short review indicates that in many species of frogs, females choose their mates. In other species, either the role of female mate choice might not be important or it might be difficult to elucidate because of the prevalence of scramble competition among males. Certainly in most species the criteria by which females select their mates are not known.

The Role of the Advertisement Call

Although most male frogs produce a variety of calls, they usually produce a specific call type that seems to play some role in long-range advertisement (Bogert 1960); this vocalization has been classified as a mating call because of its presumed function of attracting females. Wells (1977a) proposed the term *advertisement call* because many of the "mating calls" are used in male-male interactions as well as in attracting females. The term *advertisement call* appears to have gained acceptance. The advertisement call is the most conspicuous behavior of anurans. Its most obvious function, that it enables females to locate mates, has been established for a number of species (e.g., *Bufo speciosus*, Axtell 1958; *Hyla cinerea*, Blair 1958a; Gerhardt 1974, 1982; *Hyla versicolor*, Littlejohn 1958; Gerhardt 1978b).

Most studies of the advertisement call have been directed at elucidating the role of the call as a species-isolating mechanism. Closely related species almost always have calls that differ in at least one characteristic (e.g., *Bufo*, Blair 1956; *Crinia*, Littlejohn 1959; *Pseudacris*, Mecham 1961). Playback experiments, in which females are given a choice between speakers broadcasting either a conspecific or a heterospecific call, have demonstrated that females are attracted preferentially to the conspecific call (e.g., Blair and Littlejohn 1959; Littlejohn 1960; Michaud 1962; Littlejohn and Loftus-Hills 1968).

An investigation of how a male frog's sexual display influences mate choice by a female must eventually concern itself with how females perceive the information in the call; therefore, a quick review of frog hearing is in order. The investigations by Capranica and his colleagues (see reviews by Capranica 1976a, 1976b, 1977) have shed considerable light on the neurophysiological basis of acoustic processing by anurans.

Birds and mammals hear a greater range of frequencies than do frogs. To make biological sense out of these sounds, much of the decoding of acoustic information by birds and mammals takes place in the central nervous system. In contrast, the frog's peripheral auditory system, especially the inner ear, is sensitive to a narrower range of frequencies, but a range that contains the biologically most important sounds for the frog—its species-specific advertisement call.

Not only are frogs sensitive to a narrow and biologically meaningful range of frequencies, but within that range of frequencies, the inner ear tends to be "tuned" to those frequencies in the species advertisement call that contain the most energy. Unlike other terrestrial vertebrates, which possess a single organ in the inner ear for receiving acoustic information (e.g., the cochlea in mammals), frogs possess two distinct inner-ear organs: the basilar papilla and the amphibian papilla. Each papilla is innervated by a population of nerve fibers from the eighth cranial nerve. These nerve fibers couple the peripheral auditory system to the central nervous system. Each of these fibers has a distinct auditory tuning curve—some frequencies elicit a response from the fiber at a lower intensity than do other frequencies. The auditory fibers from the basilar papilla are sensitive to higher frequencies (for example, 1400 Hz in the bullfrog), while for most species the fibers from the amphibian papilla are maximally sensitive to low and mid frequencies (200 Hz and 700 Hz, respectively, in the bullfrog). The frequency sensitivities of the inner ear tend to match the spectral characteristics of the conspecific advertisement call. In some species, the low and high frequency–sensitive fibers match the advertisement call (e.g., bullfrogs, Feng, Narins, and Capranica 1975; fig. 1.1), while in others the middle and high frequency fibers (e.g., green treefrogs, Gerhardt 1974) or only the high frequency fibers (e.g., spadefoot toads, Capranica and Moffat 1975) match the call. The peripheral filtering of sounds, due to the selective frequency sensitivities of the amphibian and basilar papillae, greatly contributes to species recognition. Therefore, much of the information in the call is decoded in the peripheral auditory system. However, convergence of information from the papillae in the central nervous system, especially in the auditory thalamic region (Mudry, Constantine-Paton, and Capranica 1977; Walkowiak 1980) further processes species-specific signals.

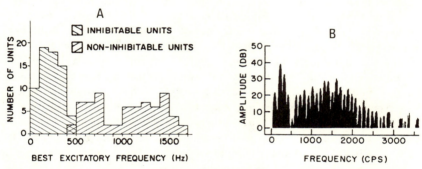

Figure 1.1 *A*, representation of the tuning curves of populations of fibers from eighth nerve innervating the amphibian and basilar papillae of the bullfrog. *B*, power spectrum (frequency versus relative amplitude) of the advertisement call of the bullfrog (redrawn from Capranica 1977).

The advertisement call not only differs among species, it also differs among populations of the same species. A number of researchers have suggested that differences among calls of closely related species are more pronounced in zones of sympatry—that is, where both species occur (e.g., Blair 1955, 1958b; Mecham 1961; Littlejohn 1965, 1969). For example, Littlejohn and Loftus-Hills (1968) have demonstrated that the advertisement calls of two species of *Hyla* (*H. ewingi* and *H. verreauxi*) are effective species-isolating mechanisms in the zone of sympatry, but these calls would not prevent mating between individuals of the two species in allopatric (i.e., spatially isolated) populations. They also have shown that female *H. verreauxi* from sympatric zones prefer calls from their own populations as opposed to conspecific calls from allopatric populations.

Capranica, Frishkopf, and Nevo (1973) demonstrated differences among populations in the spectral energy distribution of calls of the cricket frog, *Acris crepitans*. They also showed that the frequency sensitivity of the auditory system varied geographically and matched the spectral energy in the local calls. These results suggest that female cricket frogs could discriminate among calls of different populations, not necessarily that females have been selected to mate preferentially with local conspecifics (but see Shields 1982). These results could be explained by other factors that influence the male call and that, consequently, then result in a change in the frequency tuning of the female's auditory system.

Some data demonstrate that variation in the advertisement call might result in female discrimination among males within a population. Whitney and Krebs (1975) and Fellers (1979a, 1979b) have shown that females preferentially select mates on the basis of the advertisement call. This discrimination seems to be based on the total call energy received by the female, due either to differences in the number of calls (Whitney and Krebs) or the transmission qualities of the calling site (Fellers) of the male. Wells and Schwartz (1982) also found that calls from some calling sites of male glass frogs (*Centrolenella fleischmanni*) were transmitted better than calls from other sites. They suggested that this result might explain the relationship between height of the calling site and male mating success reported by Greer and Wells (1980).

Female choice may further be based on a relationship between characteristics of the male and characteristics of his call. Neurophysiological evidence suggests that female bullfrogs might be attracted preferentially to the calls of larger males because of differences among males in the spectral properties of the call. Capranica (1965) showed that the lower frequency peak of 200 Hz to 300 Hz in the advertisement call of larger bullfrogs did not occur in calls of smaller bullfrogs, which had a frequency peak at 700 Hz to 800 Hz. This peak eventually shifts to 200 Hz to 300 Hz as the frog matures and increases in size. Vocalizations were elicited from larger male bullfrogs in response to the natural calls of other larger males, or to synthetic stimuli with frequency peaks similar to that of the natural call. However, the calls of smaller bullfrogs, as well as synthetic calls lacking a 200 Hz to 300 Hz peak, did not elicit calling from larger males. The frequency sensitivity of the bullfrog's auditory system is closely matched to the power spectrum of the call of larger males (fig. 1.1). In fact, the 700 Hz to 800 Hz peak in the call of smaller males excites the fibers of the amphibian papilla that are sensitive to middle frequencies, but actually inhibits the neural response from the low frequency–sensitive fibers (Capranica and Moffat 1980). If the auditory system of the female has the same response properties, females should respond preferentially to the call of a larger male rather than a smaller male. This result is consistent with Howard's (1978a) demonstration that in nature, female bullfrogs select larger males as mates.

Gerhardt's (1974, 1982) study of species recognition in *Hyla cinerea* also demonstrated that under some conditions females were attracted preferentially by a subset of the calls in the population. The spectral energy of the species advertisement call has two peaks, at about 900 Hz and 3000 Hz. The range of frequencies for the lower peak that females preferred was considerably narrower than the frequency band in the calls of all advertising males.

Although Davies and Halliday (1978) did not examine the role of the advertisement call in female mate choice, they presented fascinating evidence on how the call of the toad *Bufo bufo* provides information about the individual's size and how this information is used in male-male interactions. Unpaired males attempt to displace mated males from the females they clasp. Fighting success is size dependent: larger males are more successful in displacing mated males. A male's vocalization indicates his size; because a significant negative correlation exists between a male's size and his call frequency, larger males have lower-pitched calls. Using playback experiments in which male size was indicated by the frequency of the call, Davies and Halliday showed that the behavior of the unmated male is influenced by the vocalization of the male he attempts to displace. Unmated males were more likely to attempt to displace smaller males. Ramer, Jennson, and Hurst (1983) also showed that in green frogs the size of the male producing the call influences the behavior of the responding males, and Arak (1983b) has demonstrated a similar phenomenon in natterjack toads.

The anuran advertisement call permits species recognition, and through the mechanism of female mate choice based on call characteristics, acts as a premating species-isolating mechanism. This is one of the better-demonstrated facts in amphibian biology. The advertisement call of some species varies locally, and female-choice experiments demonstrate that females can be attracted preferentially to the local call. Neurophysiological studies have elucidated the sensory basis for population-biased call preference. Other studies have demonstrated that differences in the advertisement call among males within a population are biologically meaningful, since the call affects the male's reproductive success. Some female anurans prefer calls of larger as opposed to smaller males, and females of other species prefer a subset of calls within the population. Individual differences in the call also provide information about the caller which is utilized in male-male conflicts.

Thus the anuran advertisement call performs functions other than that of facilitating mating between members of the same species. Although the call can only result in species isolation through female mate choice, little discussion of how the call influences mate choice below the species level has occurred. A number of factors might influence male mating success (e.g., male persistence at the breeding site), and the importance of these various factors are considered in this study of the túngara frog. However, a major objective of this research is to determine how sexual selection by female choice might influence the evolution of anuran advertisement calls.

2 The Species: *Physalaemus pustulosus*

Choosing a Species

In 1975 and 1976 I conducted a study of the territorial and reproductive behavior of the bullfrog (Ryan 1980b). At that time, territoriality had been reported in only a handful of frog species (e.g., Martof 1953; Lutz 1960; Emlen 1968b; Wiewandt 1969); and the review of anuran social behavior by Wells (1977a), and the excellent studies of bullfrog mating systems by Howard (1978a, 1978b) and Emlen (1976) were not yet published. Watching the bullfrogs, I was able to observe females in a low posture, with only their heads above the water, swimming among the calling males which were inflated, with about two-thirds of their bodies above the water's surface. As first reported by Emlen (1968b), the females swam among the calling males and appeared to select mates. A female would make physical contact with or pause in front of a calling male, who would then immediately clasp her. This type of mate choice behavior exhibited by the female bullfrogs now appears to be typical of many species with prolonged breeding seasons.

During this time I became interested in the mechanisms by which females selected mates and the factors that influenced the evolution of this behavior. I was especially interested in whether the almost constant vocalizations by the males influenced a female's choice of mates, apart from merely indicating that the males were of the correct species. I decided to investigate the role of the advertisement call in female mate choice.

In the New World, species diversity of anurans is highest in the tropics. Tropical frogs also exhibit a variety of bizarre reproductive behaviors, such as the alleged "sex role reversal" in some species of the family Dendrobatidae, the live-bearing frog *Eleutherodactylus jasperi*, or the tadpole metamorphosis that occurs in the vocal pouch of the male Darwin's frog, *Rhinoderma darwini*. But more important for studies of sexual selection, many tropical frogs have long breeding seasons, thus permitting data collection over a longer period than would be possible in the Temperate Zone. The tropics appeared to be the best place for a study of anuran sexual selection.

A potential subject for my study had to meet several criteria. In species with a long breeding season, female mate choice seems to play a more obvious role in influencing male mating success, and as just mentioned, these species provide a better opportunity for collecting large amounts of data. To avoid the problem of confounding characteristics of the male and the territory when examining the criteria by which females chose mates, I hoped to study a species in which males did not defend resources used by the females—an anuran analogue to a lekking bird. Although a number of field studies of tropical frogs have been conducted, most of these have been of a taxonomic or zoogeographic nature—and rightfully so, since plant and animal species are disappearing from the tropics at an astounding rate. Some studies do report aspects of the frog's natural history, but often not in sufficient detail to evaluate the potential of a species as a study animal.

Based on the limited natural history observations provided by Duellman (1970), the red-eyed treefrog, *Agalychnis callidryas* (Hylidae), appeared to fulfill the necessary criteria. Males produced calls with a varying number of notes while perched in trees above shallow bodies of water. According to previous reports, females approach and are clasped by calling males, and the pair then moves to the end of a branch to deposit their eggs over the water. The animals breed throughout the wet season, females appear to select mates, and males do not defend resources. The red-eyed treefrog appeared to be a promising choice.

In June 1978 I arrived at the Smithsonian Tropical Research Institute's field station on Barro Colorado Island, which is located in the middle of Gatun Lake in the Panama Canal, Republic of Panama (in 1978, however, this area was still the Canal Zone; fig. 2.1). The red-eyed treefrog proved to be an incredibly beautiful species (fig. 2.2). It has a green ground color that changes from brilliant to dull depending on the animal's behavior, yellow markings on the flank, and huge, bright red eyes. The frogs also emit a distinctive odor which might play some role in mate attraction, but I have no data to substantiate this idea.

After surveying several populations on the island, I found that the population in the laboratory clearing, beyond Kodak House, was the most dense. I began the study by marking males with waistbands, as described by Emlen (1968a), and I hoped to witness matings. Several problems developed immediately. The first was that although males do vocally advertise for mates, they only call infrequently, making it difficult to obtain recordings from most males. In fact, in this species visual cues are likely to be very important in mate attraction. This difficulty would not have been insurmountable, but there was a second problem. Much of the calling by males and the mate selection by females took place high in the canopy, sometimes more than 20 meters above the ground. Males began calling in the canopy around dusk and eventually would move down to within 4 meters of the ground, but many of the males already had mates when they descended from the tree tops. Thus observing pair formation was difficult. As I would gaze into the nighttime

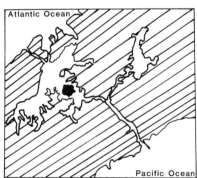

Figure 2.1 *A*, photograph of the northeast section of the island which shows the buildings in the laboratory clearing. *B*, map of Barro Colorado Island, Panama, and surrounding areas (redrawn from Karr 1982).

canopy, trying to discern the form of one of the treefrogs, there was always a great commotion at my feet. It was the cacophony of a túngara frog chorus.

. . . and Changing My Mind

To study animal behavior in the field, one needs much fortitude. For one reason or another, studies rarely proceed as outlined in a grant proposal; usually a number of unforseen difficulties must be circumvented. The many months of fruitless observation before Jane Goodall was first able to continuously observe a single individual in her study of chimpanzee behavior (Lawick-Goodall 1971) is one of the best of many examples of a researcher's

Figure 2.2 The red-eyed tree frog, Agalychnis callidryas.

early frustration. Such extreme patience and fortitude may produce great benefits later on. However, a time may come when a researcher must decide that for conducting a particular study, a certain species is not appropriate, after all. Science tends to popularize the Jane Goodall studies, but probably more numerous cases exist in which dedicated and intense research paid few or no dividends because of an initial error in choosing a species to study.

I decided that a study of the red-eyed treefrog would not provide enough data to permit a rigorous investigation of sexual selection. The túngara frog, however, appeared to be a good candidate, for it met many of the criteria I previously had established. Furthermore, Stanley Rand of the Smithsonian Tropical Research Institute had investigated the complex vocal repertoire of the species in the mid-1960s, and his data showed that female choice could be influenced by call complexity. Rand had no immediate plans to continue his studies of túngara frogs and encouraged me to conduct my study of sexual selection on this species. So I did.

Physalaemus Taxonomy and Zoogeography

The túngara frog, *Physalaemus pustulosus,* is a small, brownish-colored, fairly nondescript frog (fig. 2.3). Males average about 30 millimeters in length from the tip of their snouts to the end of their vents. Females are slightly larger. The frog's skin is pustular, thus the name *pustulosus,* and these distinctive bumps on the skin give the frog a toadlike appearance. In fact, Panamanians refer to them as toads (*sapos*) rather than frogs (*ranas*).

The túngara frog is a member of the large, cosmopolitan family Leptodactylidae. This family consists of 57 genera with 650 species, and is distributed throughout much of the Western Hemisphere (from Texas to Argentina),

Figure 2.3 A calling male túngara frog.

the Australo-Papuan region, and southern Africa (Lynch 1971; fig. 2.4A).

Physalaemus is one of 10 genera in the subfamily Leptodactylinae. The subfamily is characterized by a bony style or ossified plate in the sternum, in contrast to the cartilagenous sternum of the other leptodactylids. A foam nest is typical, though not diagnostic, of the subfamily and has been important in the species' zoogeography, as will be discussed shortly. Members of the subfamily are found from southern Texas to Chile (fig. 2.4B). They are restricted to the lowlands, with the exception of *Pleurodema*, which can be found in the Andean highlands (Lynch 1971). Noble ([1931] 1954) suggested that *Physalaemus* was the base genus of the subfamily, but Lynch (1971) considers *Pleurodema* most similar to the primitive leptodactyline stock. Nevertheless, these two genera appear to be closely related. Heyer (1975) provides the most current analysis of intergeneric relationships of the family Leptodactylidae, and a summary of his results are shown in figure 2.5.

The genus *Physalaemus* is defined by a number of osteological characteristics, the most striking of which is the location of the muscle insertion on

The Species: *Physalaemus pustulosus*   29

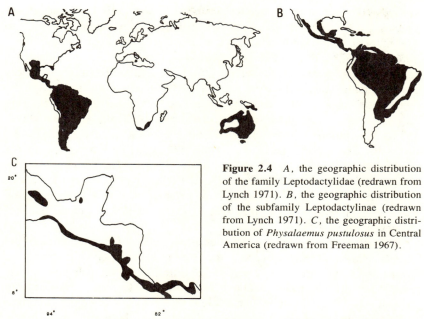

Figure 2.4 *A*, the geographic distribution of the family Leptodactylidae (redrawn from Lynch 1971). *B*, the geographic distribution of the subfamily Leptodactylinae (redrawn from Lynch 1971). *C*, the geographic distribution of *Physalaemus pustulosus* in Central America (redrawn from Freeman 1967).

the hyoid apparatus. All members of the genus exhibit axillary amplexus, in which the male grabs the female from the top and under the front legs (see fig. 3.8). *Physalaemus* lays numerous eggs in a foam nest and has pond-type larvae (Lynch 1971). The genus is distributed throughout the lowlands from Mexico to northern Argentina (fig. 2.4B). There are 35 species of *Physalaemus* which Lynch (1970) divides into 4 species groups, based mostly on

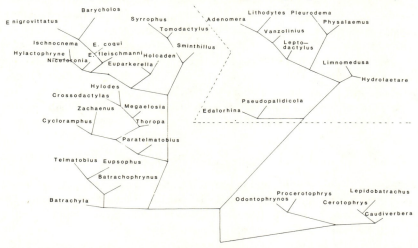

Figure 2.5 Predicted phylogenetic relationships of genera of the family Leptodactylidae. Those genera above and to the right of the dashed lines are members of the subfamily Leptodactylinae (redrawn from Heyer 1975).

skeletal characteristics. *P. pustulosus* is one of 7 species in the *P. pustulosus* species group. The species group is distributed from southern Mexico to Colombia and into Venezuela (fig. 2.4B).

Savage (1973) presents the most detailed analysis of anuran biogeographic patterns. He suggests that the leptodactylids were differentiated in situ from the liopelmatid stock (which are in evidence from at least the Triassic) in the temperate forests of South America during the Cretaceous. A remnant of the original stock is still found in the beech forests of South America (Heyer 1975). A radiation occurred and two stocks became adapted to drying conditions: the ceratophrines and the leptodactylines. The primary adaptation of the leptodactylines was to the Neotropical Tertiary Geoflora, and it has resulted in 6 genera with 11 species (Heyer 1975). The secondary adaptation, which includes *Physalaemus*, was into the savannas. This latter radiation was much more successful and is now represented by 4 genera with 85 species (Heyer 1975).

The foam nest was of primary importance in allowing the radiation of the leptodactylines into the savannas, because the nest preserves the moisture around the eggs during short periods of little rainfall. Lynch (1971) suggested that the foam nest evolved during the Cretaceous in response to climatic variations, thus allowing the leptodactylines to breed in more xeric environments. Heyer (1975, and references therein), however, suggests that the foam nest had its evolutionary origin in the wet forests; evidence that the foam nest does have some adaptive value in the wet forest is currently exhibited by the leptodactylid genus *Adenomera*. According to Heyer, the foam nest served as a preadaptation for frogs that later invaded the savannas in the Miocene and late Pliocene.

During the move into the savannas, the leptodactylines spread throughout South America and are now also found in the West Indies and in Central America. *P. pustulosus* is the only member of the genus found in Central America, and, like the other leptodactylines there, it appears to be a recent arrival. Savage (1966, 1982) suggests that *P. pustulosus* probably evolved in the xeric habitats of Colombia and Venezuela and then migrated into Central America after the reestablishment of the Panamanian isthmus in the Pliocene. Today, *P. pustulosus* is found from southern Mexico south to Colombia, and extends east into Venezuela.

Previous Natural History Studies of *Physalaemus*

A number of field studies and observations of *Physalaemus* natural history have been conducted. Most notable is the work of Barrio on the Argentinian members of the genus. He investigated the morphological, ecological, and behavioral relationships of the following species of *Physalaemus*: *P. albonatus*, *P. barbouri*, *P. bilogonigerus*, *P. cicada*, *P. cuvieri*, *P. fernandezae*, *P. gracilis*, *P. henseli*, *P. jordanensis*, *P. riograndensis*, and *P.*

santafecinus (Barrio 1953, 1964a, 1965, 1967). These studies are often concerned with reproductive isolation among *Physalaemus*, and emphasize the role of the advertisement call as a premating isolating mechanism in *Physalaemus* as well as in other anurans (e.g., Barrio 1964b, 1966). His papers also provide an excellent general review of the natural history of South Temperate Zone *Physalaemus*.

Although they lack the detail found in Barrio's studies, Milstead (1960, *P. bilogonigerus, P. fuscomaculatus, P. gracilis, P. henseli* and *P. riograndensis*), Gallardo (1970, *P. fernandezae*), Duellman (1978, *P. petersi*), and Cardosa (1981, *P. cuvieri*), also report observations of *Physalaemus* breeding behavior. Most of the other published field reports of *Physalaemus* relate to collections during various expeditions.

P. pustulosus is extremely common in certain areas of South and Central America, but few studies have been conducted in the field of their behavior or ecology. Brattstrom and Yarnell (1968) studied chorus organization of several *P. pustulosus* in an aquarium. Their conclusions about dominance hierarchies and male mating success may not apply to the species' behavior under natural conditions, and are discussed in chapter 1. Heyer and Rand (1977) described the nest-building behavior of *P. pustulosus*, and Sexton and Ortleb (1966) demonstrated preferential nest-site selection by mated pairs. Rivero and Esteves (1969) reported some observations of *P. pustulosus* reproductive behavior, and Villa (1972) presented a brief but detailed description of the calling and nesting behavior of this species in Nicaragua. There are also numerous natural history notes that add little to our knowledge of the species.

In a laboratory study, Davidson and Hough (1969) described the synchronous-asynchronous pattern of oogenesis in *P. pustulosus* and related it to the possible influence of environmental factors on the species' breeding pattern. They also presented detailed methods on maintaining a captive breeding colony of *P. pustulosus*.

Summary

I decided that a study of sexual selection by female choice would be more successful if conducted in the tropics because many of the frog species there have long breeding seasons, which would allow the time necessary to amass a sufficient data base. Also, in these species female mate choice is more obvious. Although the red-eyed tree frog initially seemed to be a promising subject, much of its behavior occurred in the canopy where it could not be observed. My initial observations in the field, and Rand's previous studies, suggested that the túngara frog was an ideal subject. Previous natural history reports of this species detailed some aspects of their ecology and reproductive behavior—further suggesting that this species was an appropriate choice.

3 Reproductive Behavior

The túngara frog, *Physalaemus pustulosus*, may be the most abundant frog in Panama; it is certainly the most conspicuous. Túngara frogs are found in a variety of habitats, usually in areas characterized by second growth, fields, or recent disturbances. They breed in shallow, slow-moving, or stagnant bodies of water, ranging anywhere in size from the water-filled hoofprint of a mammal to a large flooded field. These bodies of water usually are temporary, but the foam nest maintains a moist environment around the eggs and reduces the danger of desiccation if the breeding site dries out before the next rainfall. The frogs have also invaded more urban and suburban areas. At the onset of the wet season, at least one túngara frog seems to inhabit every gutter and ditch in and around Panama City. The males are adept at finding these ephemeral bodies of water. For example, a drinking glass left in the laboratory clearing on Barro Colorado Island, after a party one rainy night, contained the foam nest of a túngara frog the following morning!

The Breeding Season and Patterns of Oogenesis

The túngara frog breeds during every month of the year on Barro Colorado Island, but most breeding takes place during the wet season—May to December. The amount of breeding is not constant during the wet season. Males do not call during heavy rains, and the nights immediately after heavy rains are characterized by the most intense breeding. Still, how environmental factors influence this species' pattern of breeding is not clear. Figure 3.1 shows the number of calling males and nesting females at Kodak Pond from June to December 1979, as well as some corresponding environmental conditions. No obvious relationship seems to exist between patterns of mating and environmental factors.

On Barro Colorado Island, the frequent rains during the wet season drop an average of 262 centimeters of water on the island every year: 93% of the rain falls in the wet season (Rand and Rand 1982). However, even during the wet season there are periods of little or no rainfall (fig. 3.1). Thus for an animal that relies on standing water for successful development of its offspring, a degree of unpredictability is introduced into the determination of time of breeding. However, female túngara frogs possess some physiological adaptations that help them deal with this environmental unpredictability. Davidson and Hough (1969) investigated the pattern of the development and yolking of eggs (i.e., oogenesis) of the túngara frog, and suggested that females exhibit a pattern that enables them to maximize their reproductive output during times when favorable breeding conditions are unpredictable. Oocytes undergo a series of 6 developmental stages prior to ovulation. The formation of yolk bodies occurs at stage 6, just prior to ovulation.

Davidson and Hough suggested that it would be advantageous for a female that breeds in temporary bodies of water to have as many eggs as possible available for deposition when favorable breeding conditions occur. One limitation on the number of eggs produced is space, because only a certain number of yolk-filled oocytes (stage 6) can fit in the ovary. For an animal that breeds at frequent but unpredictable intervals, it would be advantageous to have a set of oocytes ready to pass from stage 5 to stage 6 immediately after ovulation of the previous set. Thus a pattern of asynchronous oogenesis is predicted—that is, oocytes should be found in similar numbers in all stages. During the dry season when females rarely breed, it is less important for a female to have her oocytes in staggered stages of development. At this time, maximizing the number of eggs deposited in the first clutch of the wet season might even be advantageous. Therefore, during the dry season no advantage for asynchronous oogenesis exists, and a female might even benefit by maximizing the number of oocytes in stage 6 at the expense of oocytes in other stages. A pattern of synchronous oogenesis might be expected during the dry season—that is, most oocytes should be in the same stage of development.

By injecting an isotope into gravid females in the laboratory, Davidson and Hough followed the development of oocytes—the first time this procedure was attempted with an amphibian. They then simulated wet and dry seasons in the lab and described the pattern of oogenesis under each regime. Their model of oogenesis is depicted in figure 3.2. They found that during the dry season the development of oocytes tends to be synchronous—most of the oocytes are in stage 6—which results largely from the restriction of oocytes advancing from stage 3 to stage 4. Consequently, the clutch size at the end of the dry season, which is the onset of the wet season, is larger than the clutch sizes during the wet season (450 during the dry season versus 80 to 300 during the wet season). As the wet season progresses, the number of oocytes in stage 3 to stage 6 begin to equilibrate. The result of this asynchronous pattern of oogenesis is that females have smaller clutches, but they are able to deposit these clutches at more frequent intervals. Therefore, the synchronous-asynchronous pattern of oogenesis is especially suited to the environmental conditions under which túngara frogs breed.

34   Chapter Three

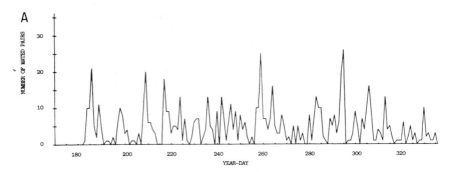

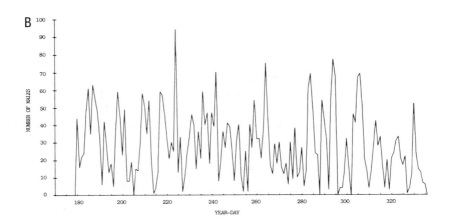

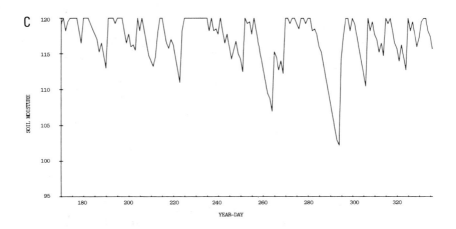

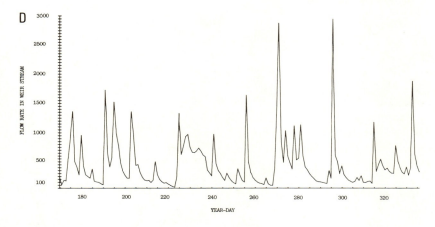

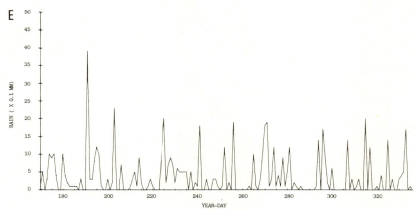

Figure 3.1 The number of male and female *Physalaemus pustulosus* at Kodak Pond from 30 June to 2 December 1979, along with corresponding environmental conditions. Stream flow was recorded in the nearby Lutz Stream.

Calling and Mate Choice Behavior

Túngara frogs are especially abundant on Barro Colorado Island and breed in a variety of habitats there, predominantly in shallow pools along streams, puddles along trails, and in a number of places in and around the laboratory clearing. Most of my observations of *P. pustulosus* were conducted at a small cement pool (1 meter by 2 meters) behind Kodak House in the laboratory clearing on Barro Colorado Island (fig. 3.3). Túngara frogs have been breeding in this pond for at least 15 years—since the mid-1960s when it was constructed by Rand to study the frogs' vocal behavior. However, I have observed this species breeding in a variety of other habitats, and I have not observed any significant differences in their behavior among breeding sites.

36   Chapter Three

The túngara frog is especially conspicuous because of its unique vocalization. The advertisement call has two components, a "whine" and a "chuck"; the call resembles the sounds produced by some "star war" video games. The call always contains a whine and can contain from 0 to 6 chucks (fig. 3.4; the structure and function of the advertisement call are discussed in detail in chap. 5). In fact, the Panamanian word for *P. pustulosus* is onomatopoetic: the *túng* resembles the whine and the *ara* the chucks. When the first syllable is stressed and the second syllable is pronounced very rapidly, *túngara* sounds remarkably like a call with a whine and 2 chucks. *Túngara* is also the common name of *P. pustulosus* in Nicaragua (Villa 1972), but they are called *púngala* frogs in Trinidad.

The nightly breeding activities of the túngara frog begin at dusk when males first appear at the pond and immediately begin to produce whines, or whines plus chucks. The amount of calling slowly increases during the night

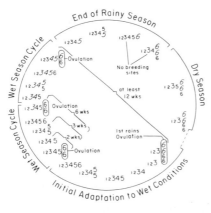

Figure 3.2 *Left:* A schematic representation of the model of patterns of oogenesis of *Physalaemus pustulosus* proposed by Davidson and Hough (from Davidson and Hough 1969).

Figure 3.3 *Below:* Kodak Pond.

until some time between 2000 and 2200 hours, and then decreases substantially around midnight (fig. 3.5). The males are usually located around the perimeter of the pool facing away from the center (fig. 3.6). Males that have been calling are found in an inflated position, their lungs filled with air ready to be expelled for the next call. Males that have not been calling are in a deflated position. Calling males maintain a fixed distance between themselves and their nearest calling neighbor by exchanging "mew" calls (fig. 3.7) and by clasping any frog, of any species, that comes within about 5 to 10 centimeters.

Males spend most of the time from 1800 to 2400 hours calling, although they cease calling when disturbed, especially by the presence of predators

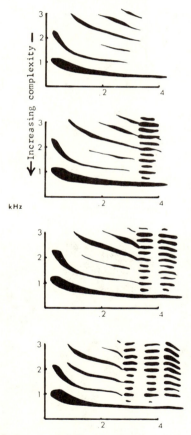

Figure 3.4 A sketch of sonograms of the advertisement call complexity series of *Physalaemus pustulosus*. The calls increase in complexity from top to bottom, with a whine only at top and a whine plus 3 chucks at bottom. Time is on the horizontal axis and frequency is on the vertical axis (from Rand and Ryan 1981).

38   Chapter Three

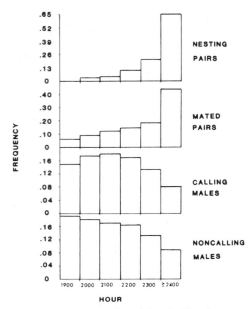

Figure 3.5 The distribution of various male and female *Physalaemus pustulosus* behaviors throughout the night at Kodak Pond in 1979. Time of night is represented on the horizontal axis. The proportion of the total number of times that a behavior was observed in each hour is on the vertical axis. For example, 20% of the mated pairs were observed in the time period from 2300 to 2400 hours (from Ryan 1983b).

Figure 3.6 Calling male *Physalaemus pustulosus* at Kodak Pond; a tag used to identify individuals can be seen on the male's back.

(chap. 7). When a male begins to call, his call repetition rate is variable. But after calling for a short time, most males call at what seems to be a maximum call repetition rate of about 1 call every 2 seconds. Like most frog calls, the *P. pustulosus* call is loud, about 90 dB SPL at 50 centimeters from the frog. A decibel (dB) is a logarithmic comparison of sound pressures. The pressure of the sound being measured is compared to some reference sound pressure level. Because the comparison is logarithmic and not linear, a sound that is 6 dB more intense than the reference sound pressure produces twice as much pressure. SPL indicates that the reference sound pressure is .0002 dynes/square centimeter (0 dB), which is the sound pressure at which humans can barely detect a 1000 Hz sound. For comparison, a heavy truck passing by at a distance of 15 meters registers about 90 dB SPL (Rossing 1982).

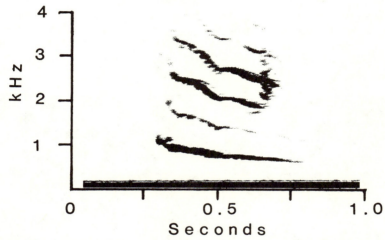

Figure 3.7 A sonogram of a "mew" call of *Physalaemus pustulosus*. Time is on the horizontal axis and frequency is on the vertical axis.

Male túngara frogs probably call more in one night than any other frog that has been studied. We wanted to know precisely how many calls a túngara frog produces in a single night (chap. 7). To do this, near a chorus we placed a single male in a plastic bucket with a screened top. On top of the screen we placed a microphone that, through a tape recorder, relayed the male's call to a chart level recorder. The paper of the chart level recorder moved at a constant speed, and every time the male called a pen marked the chart paper. This procedure allowed us to count the total number of calls produced by the experimental male in the bucket. The calls from the chorus also were recorded on the chart paper, but they were easy to distinguish from the experimental male because the chorus was much farther from the microphone. We found that the experimental male started and stopped calling with the chorus. Thus the bucket did not seem to influence the number of calls produced by the experimental male.

Chapter Three

Table 3.1 Calls Produced per Night by Male *Physalaemus pustulosus*

Hour	1	2	3	4	5
1900	408	1196	960	899	593
2000	854	1229	720	1757	825
2100	1190	1011	705	1602	651
2200	808	1282	588	1085	339
2300	731	1538	417	776	82
2400	217	988	409	127	0
0100	393	0	0	0	0
0200	0	0	0	0	0
Total	4601	7244	3830	6246	2490

Note: Each number in the heading represents a single frog.

Five males were monitored from 1900 to 0200 hours. One male called during all 7 hours, 4 males discontinued calling during hour 6, and 1 stopped after 5 hours (table 3.1). In the hours in which males called, the average number of calls was 830.6 calls/hour (standard deviation = 247.7). Given a maximum call repetition rate of 1 call every 2 seconds, if a male called

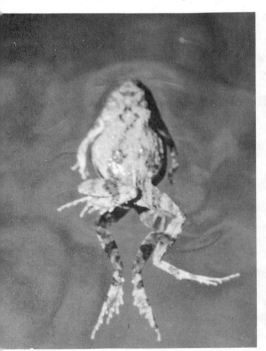

Figure 3.8 *A*, male and female *Physalaemus pustulosus* in amplexus; *B*, mated pair as they swim through the water.

continuously for an entire hour the maximum number of calls produced would be 1800. The average number of calls per hour produced in these experiments was 46% of the maximum.

Females usually come to the breeding site only on the night they mate. They often sit on the shoreline in front of calling males, or enter the water swimming among the calling males, sometimes stopping 10 or 20 centimeters in front of a male for several minutes. The males do not actively search for females, but they remain at their calling sites clasping any frog that approaches within a few centimeters (fig. 3.8). If the intruding frog is a conspecific male, he usually gives a release call when clasped (fig. 3.9) and is then promptly released. Sometimes the intruding male is not released, and a fight ensues with the clasped frogs rolling through the water. On a few occasions a male died during one of these prolonged wrestling bouts. Sometimes male túngara frogs are found clasping much larger toads (*Bufo typhonius;* fig. 3.10). Whether these male toads failed to give a release call,

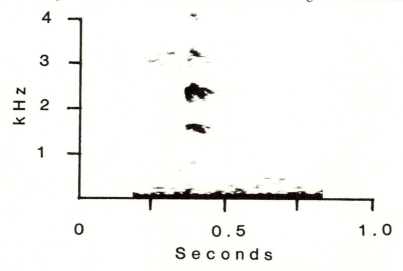

Figure 3.9 *Above:* A sonogram of a release call. Time is on the horizontal axis and frequency is on the vertical axis.

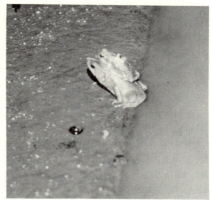

Figure 3.10 *Right:* A male *Physalaemus pustulosus* clasping a male *Bufo typhonius*.

or whether the túngara frogs were merely not responsive to their calls is not known. However, Noble and Aronson (1942) pointed out that the stimulus causing release after the release call may not be an auditory cue, but probably is the physical vibration in the animal's flank caused by the call. They suggested that clasping males feel this vibration, and that this is the stimulus that causes the release. If so, then when a male túngara frog clasps a *B. typhonius*, the túngara might not be able to perceive the vibratory stimulus necessary to elicit release because his forelegs barely reach around the back of the male toad.

Female túngara frogs initiate matings in much the same way as do female bullfrogs and some other prolonged breeding anurans. Females approach and either stop directly in front of or make physical contact with calling males, which then clasp them; females have been observed to pass by some calling males en route to others (Emlen 1968b, 1976). Females of both species also appear able to move among the chorusing frogs with minimal interference from other males, probably because males are not actively searching for mates.

On many occasions I observed an unmated female sitting in front of a calling male and then quickly swimming into the calling male, at which time she was clasped. Fewer observations were made of extensive movements of the female within the chorus prior to mating, but some of these observations are shown in figure 3.11A–C. The female whose behavior is shown in figure 3.11A was sitting in the water near the shore about ⅓ of a meter from a calling male. She sat near the male for about 3 minutes and then quickly turned and swam across the pool to the opposite shore, where she again positioned herself about ⅓ of a meter from another calling male. After about 5 minutes she circled around on the shore and jumped into the water. She then made contact with the calling male and immediately was clasped by him. Figure 3.11B describes the movements of another female prior to mating. This female sat on the shore in front of several calling males for about 4 minutes, and then swam past these males and sat in front of another male for about 1 minute before swimming to the nearby shore. As the female approached the calling male on this shore, he gave a mew call, as if the female were an intruding male. But he quickly resumed producing the advertisement call, and the female swam into him, initiating mating.

Figure 3.11C shows the movements of a female that sat in the corner of the breeding site, moved away to other calling males, and then returned to a point near her previous location. After the female left the corner, she sat in front of a male near the center of the adjacent shore. When she attempted to move past this calling male onto the shore, he grabbed her, but the female was able quickly to wriggle free from his grasp and move onto the shore where she remained for several minutes. She then moved into the water, stopped in front of some calling males, and then continued to swim back towards the corner. During this time, a calling male had shifted his position away from the shore, and the female's movements brought her directly into this calling male. In this case it appeared that the female's movements were actually directed to the calling male on the far side of the male that clasped her. However, it is certainly easier to record the female's movements than it is to interpret her motivations.

A basic assumption of these interpretations of the females' behavior as

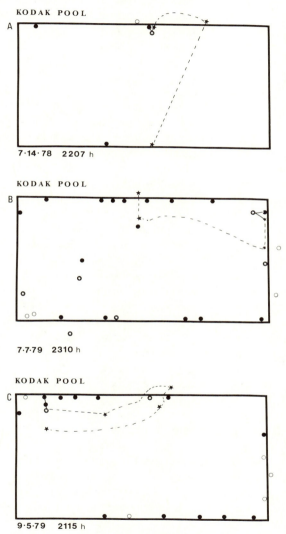

Figure 3.11 Movements of female *Physalaemus pustulosus* prior to, and at the initiation of mating. Stars represent females, solid circles represent calling males (i.e., males in the inflated position), and hollow circles represent noncalling males (i.e., males in the deflated position). A dark circle with a hollow star represents a mated pair. The dashed line traces the movement of a female. The time when the observations began and date of the observation are given on each diagram.

"choice" is that the initiation of mating is not due to random movements on the females' part, but that it is the result of some goal-directed behavior. This assumption holds whether the female is attracted to the most intense stimulus (i.e., passive choice), or whether she actively compares possible alternatives (active choice, Parker 1983; see also Halliday 1983a, 1983b). My observations of the females' movements suggest that females are sampling males before they initiate mating. This interpretation will be strongly supported by the data on male mating success, as well as by the results of the female playback experiments presented in chapters 4 and 5.

Nesting Behavior

Mating occurs throughout the night, but as figure 3.5 shows, mated pairs are most commonly observed toward the end of the night. Pairs often leave the pool immediately after clasping; therefore, many mated pairs are first sighted when making the nest. After contact between the sexes is initiated, the male clasps the female: amplexus has occurred. Only on a few occasions did I observe unmated males attempting to displace mated males from females. All of these attempts were unsuccessful. Quite often, mated pairs sit directly in front of calling males with no interference from the unmated animals. Most interference from unmated males probably occurs when pairs happen to move within the clasping range of calling males. Once when this happened, the mated male gave a release call and the mated pair was released immediately by the calling male. This behavior contrasts with that of some explosive breeders, such as the toad *Bufo bufo* (Davies and Halliday 1977, 1978, 1979), where displacement of rivals is an important determinant of male mating success. Instead, the túngara frog's behavior appears to be typical for males of species with prolonged breeding seasons (Wells 1977a). Several factors can explain the rarity of displacement behavior by male túngara frogs. In toads, the size difference between the sexes is much larger than in túngara frogs. A male's success in displacing a mated male is dependent on his size, but it is also influenced by the size difference between the male and female of the mated pair. The larger the size difference, the more difficult for the mated male to maintain his grasp. Perhaps the infrequent occurrence of displacements in túngara frogs is because the sexes are closer in size, and thus the mated male would have a better grip on the female and would be much more difficult to displace. The tendency of pairs to leave the breeding site soon after clasping might function to reduce the probability of interference from other males. But there are other possible explanations for this behavior.

When the pairs leave the breeding site they usually do not return until calling has subsided. Although the function of calling is to attract mates, females are not the only ones attracted to calling males—calls also attract predators. Therefore, mated pairs might avoid predation, as well as interference from unmated males, if they flee the breeding site during the poten-

Figure 3.12 Nest construction by a male and female *Physalaemus pustulosus*.

tially dangerous period of chorusing. Most of the mated pairs return to the breeding site around midnight, and nest construction usually begins between 2400 and 0200 hours. On most nights all breeding activity ceases before 0400 hours (fig. 3.5).

The most unusual aspect of the reproductive behavior of túngara frogs is the construction of the foam nest. Heyer and Rand (1977) have described this behavior from a series of films. The female extrudes several eggs from her cloaca as the male clasps her from the top (fig. 3.12). The male moves his

Physalaemus pustulosus foam nest construction. Numbers are number of frames, first frame in sequence diagrammed being zero.

Physalaemus pustulosus foam nest construction. Numbers are number of frames, first frame in sequence diagrammed being zero.

Figure 3.13 A diagram of foam nest construction from the side and top. The number beside each tracing represents the frame of the movie film from which the tracing was made (from Heyer and Rand 1977).

back legs forward, placing them under the female's cloaca. He takes the eggs between his back feet, moves them past his cloaca, where fertilization takes place, and then beats the eggs with several short quick strokes of his hind legs (fig. 3.13). The jelly matrix surrounding the eggs retains air and becomes foam, similar to the foam produced when chicken egg whites are beaten into a meringue. The adaptation that results in the foam nest is primarily behavioral, not physiological, since supposedly any frog that beats its eggs would have a foam nest (Heyer 1969). Among the leptodactylid foam nesters of Australia, the females beat the foam with their hands (Martin 1970).

Nest construction lasts about an hour, and the male fertilizes an average of 230 eggs. This behavior is energetically demanding, and the process visibly tires the male—the time interval between strokes tends to increase toward the end of nest construction. When nesting is completed the female usually swims out from under the male; the male lingers behind for a minute or so before he too quickly disappears into the night.

The female is responsible for locomotion of the pair once they are in amplexus. The male merely hangs on to her back as the female searches for a location to place the nest. Females will not place the nest just anywhere: a pair might spend more than two hours moving about the breeding site before nest construction begins. What is a female looking for when she searches for a nest site? Sexton and Ortleb (1966) suggest that females prefer nest sites on a vertical plane or a curved surface intersecting and perpendicular to the water, because the female needs an object to support her front legs. However, the films of Heyer and Rand clearly show that no submergent support is used by the female. These findings do not question the nest-site preference demonstrated by Sexton and Ortleb, only the reason for the preference.

My observations support the results of Sexton and Ortleb, but offer another, and perhaps a more important consideration of nest-site selection. Túngara frogs show a distinct preference for nesting communally. By that, I mean that pairs preferentially attach their nests to other fresh nests (less than 1 day old) when possible. Communal nesting behavior involves more than simple nest-site preference. As the mated pairs slowly swim around the perimeter of the pool, pairs sometimes stop and congregate on the shore line. When one pair begins constructing a nest the other pairs quickly join in (fig. 3.14). On one occasion, several pairs formed a semicircle with their cloacas facing toward the center of the circle and then began nest construction simultaneously.

As discussed in chapter 2, the foam nest seems to have played an important role in the biogeography of the leptodactylines, allowing the group to invade the savannas during either the Cretaceous (Lynch 1971) or in the Pliocene and the Miocene (Heyer 1975). During times between rains, when temporary pools might dry up, the foam nest retains moisture and assures the continued survival of the eggs and larvae. This adaptation probably allowed the lepto-

Figure 3.14 Communal nesting by *Physalaemus pustulosus*.

dactylines to exploit a new adaptive zone and to escape the larval competition in the pond-marsh environment (Heyer 1969). The foam nest also decreases predation on the eggs by keeping most of them out of the water and away from potential egg predators. Often I have observed the larger *Agalychnis callidryas* tadpoles eating *P. pustulosus* eggs from the submerged parts of the foam nest. During heavy rains, the nests break down and the eggs float on the water's surface, where they become easy prey for *A. callidryas* tadpoles and, no doubt, other predators.

The magnitude of these two benefits of the foam nest—reduced predation and reduced desiccation—probably depends on the surface-to-volume ratio of the nest, which might explain the tendency for communal nesting. For a given volume, the smaller the surface exposed to the environment, the smaller the number of eggs exposed to predation and desiccation. As the nest increases in size, the volume and surface area increase disproportionately—the volume increases to the third power and the surface area to the second power. All else being equal, larger nests should have a smaller surface-to-volume ratio and thus suffer relatively less predation and desiccation than smaller nests. A potential problem that accompanies larger nest size is adequate diffusion of oxygen to the nest's center. Given the porous nature of the foam, however, oxygen probably is not a limiting factor. Therefore, túngara frogs probably engage in communal nesting because the accumulation or aggregation of nests causes each pair's eggs to suffer less predation and desiccation than if each nest were deposited in isolation.

These observations of the species' reproductive behavior, and especially the nesting behavior, bear directly on the question of whether or not males defend territories. The answer depends on the definition of territory. Some researchers have proposed that territoriality entails site fidelity (e.g., Nice 1941). By this definition túngara frogs are not territorial, because males do not necessarily return to the same calling site on successive nights. Wilson (1975) suggested that animals might defend "floating" territories; thus, any animal that maintains a distance between itself and its neighbor would be territorial. By this definition túngara frogs are territorial because males do maintain individual distances. The important point in regard to the species' mating system is not whether the frogs are territorial per se, but whether males defend resources that are used by females. If so, then the quality of the resource needs to be considered as a potential basis for female mate choice. The observations just reported indicate that males do not defend resources used by the female in reproduction. The only resource a male could potentially defend is the nest site. However, pairs usually leave the breeding site after clasping and construct the nest upon their later return to the breeding site. They do not necessarily return to the male's previous calling site. In fact, what appears to be the most important cue in nest site selection—the presence of fresh nests—is not available when the male begins calling at dusk. Therefore, I conclude that females are not utilizing resources defended by males; we can therefore abandon the semantic debate of whether or not males are truly territorial.

Summary of Reproductive Behaviors

These observations of the túngara frog's breeding behavior have some important implications for a study of sexual selection. Males advertise vocally from the breeding site. The call certainly plays a role both in interactions among males and in attracting females. Male-male interactions result in maintenance of individual distances. A male's probability of acquiring a mate should be greater when neighboring males are kept at a distance. Females swim among the calling males with little or no interference. Eventually, a female initiates amplexus by making contact with a calling male. Thus females behave in a manner that enables them to sample mates. Males do not defend resources that are utilized by females, and the calling site does not later serve as the oviposition site. In fact, the most important criterion for a nest site—other nests—is not present when females choose mates. The only contributions made by males are sperm and assistance in nest construction. These observations strongly suggest that females select males, and because resource quality will not enter into the female's decision, female mate choice is based on the male himself.

4 Male Reproductive Success

Reproductive success is a measure of the progeny produced by an individual in the population, relative to the number of progeny produced by other individuals. Sexual selection results in some individuals having greater reproductive success than others. In túngara frogs, we know that females select mates, but we must ask if this process results in differential reproductive success among males. Since reproductive success is the primary component of fitness, the variance in reproductive success is directly related to the intensity of selection; therefore, it estimates the potential for the differential increase of a male's alleles in the population (see chap. 1). The importance of differential reproductive success in evolutionary theory should be clear. But reproductive success can be estimated at several levels, from the number of females that a male consorts with, to the number of grand offspring he sires (Howard 1979). Reproductive success is often estimated in different ways for different taxa. For example, in studies of avian mating systems a male is often given credit for offspring produced by females in his territory. Since multiple mating may occur among birds (Bray, Kennedy, and Guarino 1975), the certainty of assigning offspring to any particular male can be low. However, once the eggs hatch, the researcher often can make accurate estimates of fledgling success and recruitment of those offspring into the adult population. In some cases the reproductive success in the second, or more distant, generations can be estimated.

With few exceptions (e.g., Townsend et al. 1981), frogs' eggs are fertilized externally, which, as pointed out in the first chapter, allows the researcher to be certain of the paternity of the female's clutch. Therefore, male mating success can be measured accurately, and in frogs the variance in male mating success is probably the primary component of the variance in male reproductive success (Howard 1979, 1983). In some frogs, but not *Physalaemus*, the eggs are deposited in a clear film, allowing the number of eggs fertilized by the male as well as the eggs' hatching success to be quantified (as in bullfrogs, Howard 1978a, 1978b). However, the frog's offspring cannot be followed after hatching. Once the eggs hatch, the tadpoles disperse in the water and then further disperse as small frogs upon metamorphosis. Virtually no information about frogs is available from the time they metamorphose until the time they begin to breed. Thus at some levels, reproductive success of male frogs can be estimated more accurately than it can for other taxa, but at other levels, such as determining the number of offspring that survive to breeding age, male reproductive success cannot be estimated accurately.

The túngara frog presents both advantages and disadvantages for estimating male reproductive success. Not only is fertilization external, but mated pairs construct foam nests. The nesting behavior is both conspicuous and of long duration (about 1 hour), and the frogs will continue to construct the nest within inches of an observer. Male mating success is not confounded by multiple matings, and because nest construction is so conspicuous and of such long duration, the observer would likely not miss any mated pairs constructing a nest. Therefore, the number of matings by a male at the breeding site can be measured with virtually no error. The disadvantage is that the eggs are inside the foam nest, which makes it impossible to monitor clutch size and hatching success in the field.

Male Mating Success: Pilot Study, 1978

When I began the pilot study of the túngara frogs' mating system in 1978, the important objective was to measure male mating success. During 12 weeks, from June to August, I observed the population of *P. pustulosus* that bred in the small cement pool behind Kodak House in the laboratory clearing on Barro Colorado Island. As shown in figure 4.1, the level of breeding activity was not constant during the study. On all nights more males present than females were present. Since females usually came to the site only on the night they mated, the number of females equals the number of matings per night. And because males mated only once per night, with only one exception observed in 2 years, the number of females also represents the number of mated males.

Emlen and Oring (1977) suggested that the intensity of male competition for mates is dependent on the operational sex ratio, defined as the ratio of potentially breeding males to females at a given time. The reasoning is straightforward: the greater the skew in the operational sex ratio, the greater the potential skew in mating success. The operational sex ratio on a nightly basis at Kodak Pond was quite variable, ranging from 13.5 (27:2; males to females) to 1.8 (9:5), although there was a significant correlation between the number of males and females at the breeding site ($r = .69$, $P < .01$, $N = 48$). On most nights at least three times as many males were present as females, an average of 9.56 (standard deviation = 6.5) males to 2.04 (standard deviation = 2.5) females. Therefore, on all nights males should have

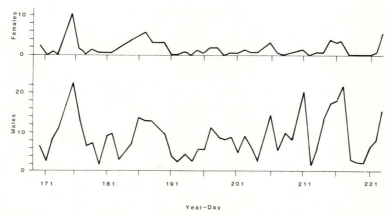

Figure 4.1 The number of male and female *Physalaemus pustulosus* at Kodak Pond from 20 June to 5 August 1978.

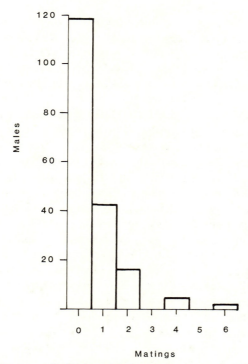

Figure 4.2 The frequency distribution of male mating success in 1978. The number of matings is on the horizontal axis and the number of males that obtained each number of matings is on the vertical axis.

been competing for mates, because not enough females were present for all males to be mated. This competition should have been quite intense, with an average of only about 20% of the males acquiring mates.

In 1978 I marked 185 males and observed 103 matings (fig. 4.2). Sixty-four percent of the males marked during the study did not mate. The mean number of matings per male was 0.57 with a variance of 0.91. These results clearly indicate significant variation in male mating success. Of course, it is not a corollary that all, or any of this variation is due to sexual selection.

Not all males were present at the breeding site on all nights; the number of nights on which a given male was present ranged from 1 to 16. The average number of nights a male was present at the breeding site was 3.27 (standard deviation = 1.75). During the study I routinely surveyed nearby túngara frog breeding sites for males that I had marked at Kodak Pond. Only rarely were marked males found at other sites. Although I did not know the location of marked males when they were away from Kodak Pond, that they regularly visited other breeding sites seems unlikely. Therefore, males probably did not breed on most nights they were absent from Kodak Pond.

A male's mating success was strongly influenced by the number of nights he spent at the breeding site ($r = .69$, $P < .01$, $N = 185$). Calling and mating is a considerable energy drain for males (chap. 8), and among other things, the ability to obtain food might influence the number of nights a male might spend at the breeding site. No data exist to test this interpretation. One factor that did not influence the time a male spent at the breeding site was his size, as the correlation between these two factors was not significant ($r = .17$, $P > .05$, $N = 185$).

The number of males in each size class is shown in figure 4.3 (see the denominator in each bar). As pointed out in chapter 1, larger males are more likely to acquire mates in several species of frogs. This was also true for *P. pustulosus*. Figure 4.3 shows two measures of male mating success as a function of male size. The first presents the proportion of successful males in each size class; success is defined as obtaining at least one mating. The second measure shows the total number of matings per size class divided by the number of males in that size class. Both the number of successful males and the proportion of matings per size class increase as a function of male size. For example, only 5 of the 19 males between 26 and 28 millimeters mated and each of the successful males mated only once, while 8 of the 14 males in the 32 to 34 millimeters size class mated a total of 13 times. Therefore, larger males are more likely to mate. Again, this is not because they spend more time at the pond since no correlation exists between size and time spent at the breeding site.

Measures of male reproductive success over the breeding season are obviously important when discussing sexual selection. However, when specifically addressing the role of female mate choice, the measures of male reproductive success over the entire season are not necessarily the most

Male Reproductive Success   53

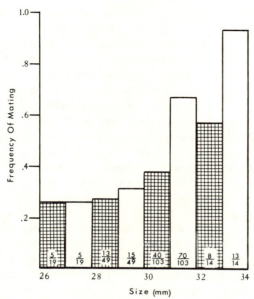

Figure 4.3 The frequency of mating as a function of male size in 1978. Male size classes are on the horizontal axis and mating frequency is on the vertical axis. The checked bars represent the frequency of successful males in each size class—that is, the number of males that obtained at least one mating. The denominator in each checked bar is the number of males in that size class, and the numerator is the number of successful males in that size class. The hollow bars show the total mating success as a function of male size. In the hollow bars the numerator is the total number of matings obtained by males in each size class. From M. J. Ryan, "Female mate choice in a neotropical frog," *Science* 209 (1980): 523–25. Copyright 1980 by the American Association for the Advancement of Science.

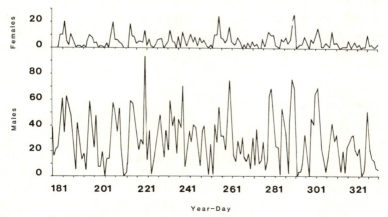

Figure 4.4 The number of males and females present at Kodak Pond from 30 June to 2 December 1979 (from Ryan 1983b).

relevant data. For example, the behavioral observations I reported indicate that female túngara frogs have the opportunity to choose mates. The data just presented show that larger males are more likely to obtain mates, and the obvious conclusion is that females choose larger males as mates. But the data also show that a male increases his chances of mating by being present at the breeding site. If larger males were more likely to be present at the breeding site (which is not the case here), then females could select males randomly with respect to size. When the data were pooled over the entire season, however, it would appear that females preferentially selected larger males as mates. Many of the studies reviewed in chapter 1 employ such a logic. But to properly address the question of female preference, one must know the sample of males from which a particular female selects her mate.

Knowing how many males or which males are sampled by a female before she finally mates is difficult. The limited observations of female movements among males prior to mating that I presented in the previous chapter (fig. 3.11) suggest that females might sample a large proportion of the males present. (Of course, the females sampling many males are the ones most likely to be seen by the observer.) Since a female usually is present only on the night she mates, I have assumed that the males present on that night represent the set of potential mates she can sample, and I use this estimate to examine female mate preference.

One prediction from the data showing size-biased male mating success over the 12-week pilot study is that females actually choose larger males. I tested whether this biased mating success is also a nightly phenomenon, and thus potentially a result of female choice. I compared the size of mated males to the size of all the males present at the breeding site on the night the former group mated. The null hypothesis of no size-biased choice by the female predicts that mated males are equally as likely to be smaller as they are to be larger than the mean size of males present on that night. Therefore, the null hypothesis predicts that of the 103 matings observed, 51.5 of the males should have been smaller and 51.5 larger than the mean male size. Table 4.1 shows that mated males usually were larger than the average male at the breeding site on the night they mated.

Summary of the Pilot Study

The behavioral observations of *P. pustulosus* reproductive behavior suggested that female choice is an important determinant of male mating success. The measure and analysis of male mating success shows that larger males are more likely to mate over the entire season, as well as on a nightly basis, which suggests that females preferentially select larger males as mates. Several predictions resulted from the pilot study. The first was that the phenomenon of size-biased female choice was a real phenomenon, and one that would be apparent in a more detailed study over a longer period of time.

Table 4.1 Size of Mated Males Relative to the Mean Size of Males Present

	$< \bar{x}$ SVL	$> \bar{x}$ SVL	Total
Observed	32	71	103
Expected	51.5	51.5	103
			$X^2 = 14.15$
			$P < 0.005$

These results also led to predictions about how females select their mates. Because males did not defend resources, it seemed likely that females selected mates on the basis of some male trait. The advertisement call is the most conspicuous character of a reproductively active male. As indicated earlier, it is well known that most female frogs use this call to select mates of the appropriate species. The call also could influence female choice among conspecific males in at least two ways. First, females might be attracted preferentially to males that call more frequently, or males that produce more intense calls; in other words, quantitative aspects of calling might influence female choice. Second, females might be attracted preferentially to calls that contain certain temporal or spectral characteristics; that is, qualitative aspects of the call might influence female choice. Most anurans rely on qualitative aspects of the call in species recognition. I was specifically interested in whether or not a differential response by females to advertisement calls explains why large males were more likely to be selected as mates. The first possibility suggests that larger males either call more or they produce more intense calls. The second possibility suggests that some call characteristic correlates with male size and that females respond to this characteristic in such a way that they are more likely to be attracted to the calls of larger males. These possibilities were considered during the next two breeding seasons.

Male Mating Success, 1979

I returned to Panama in June of 1979 for a two-and-a-half-year stay on Barro Colorado Island. One immediate goal was to monitor male mating success again, but over a longer time period, to determine if differential mating success of larger males was a general phenomenon. I did this on 152 consecutive nights at Kodak Pond from 30 June to 2 December. During this period I marked 617 males and observed 751 matings.

As in 1978, the number of males and females at the breeding site varied during the study (fig. 4.4), although the breeding population was larger in 1979 than in 1978 (a mean of 27 males versus 9.6 males). Again, more males than females were always present at the breeding site, and the numbers of males and females at the breeding site were highly correlated ($r = .72$, $P < .01$, $N = 152$). The operational sex ratio for the breeding season was 5.5 (27:4.9).

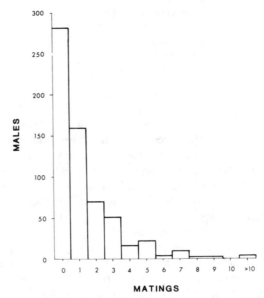

Figure 4.5 The frequency distribution of male mating success in 1979. The number of matings is on the horizontal axis and the number of males that obtained each number of matings is on the vertical axis.

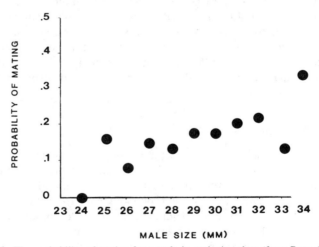

Figure 4.6 The probability of mating for a male in each size class (from Ryan 1983b).

The average number of nights on which an individual male was present at Kodak Pond was 7.2 (standard deviation = 7.1), but a large range was found in the number of nights that males were present—from 47 to 1. Males were monitored (time from first capture to last resighting) for an average of 43.0 days (standard deviation = 40.0). Therefore, the average male was at the breeding site on 16.7% of the nights that he was known to be alive. The total number of frog-nights spent at the pond by the 617 males during the 152 day study was 4456. (Each night a frog spends at the pond equals one frog-night.)

Male size did not change during the 12-week pilot study in 1978, but males did grow during the longer study in 1979. Therefore, the mating success, or other behaviors, of a male who was one size during the study could not be considered because male size changed. In determining how a male's size influenced his mating success, it was necessary to know his size on each night that he was at the breeding site.

Male size was determined in one of three ways. On the night that a male was measured, either because it was his initial capture or because he had lost his tag and was being remeasured, his size was known. If a male was present on a night after his initial capture but before some subsequent recapture, his size could easily be interpolated since his size on dates both before and after his appearance at the breeding site was known. Occasionally, some males were at the pond after their final recapture and remeasurement. Average growth rates for males of different sizes were determined. Túngara frogs, like other vertebrate ectotherms, exhibit indeterminant growth—that is, they continue to grow after reaching sexual maturity. But they also exhibit another common phenomenon of ectothermic growth (e.g., Howard 1980): growth rates decrease with size (see fig. 6.1). Using this population growth rate information, and knowing the number of days since a male's last re-measurement and his size on that day, a male's size was extrapolated for nights at the breeding site after his final measurement. In retrospect, the results of the analysis of male mating success probably would not have changed if I had used only a male's initial size, since males seldom grew more than 1 millimeter. However, the data on growth rates, and especially the variance in growth rates, will prove useful when later discussing the potential adaptive significance of female mate choice.

Not only was the breeding population larger in 1979, the average number of matings per male was higher: 1.22 (1979) versus 0.67 (1978). In part, this finding was due to the less skewed operational sex ratio in 1979. As figure 4.5 reveals, a skewed distribution of male mating success occurred again, with 45% of the marked males not mating at all. The variance in male mating success was greater in 1979 than in 1978 (3.14 versus 0.83), as was the coefficient of variation (145 [1979] versus 136 [1978]).

The amount of time a male spent at the breeding site again strongly influenced his mating success ($r = .74$, $P < .01$, $N = 617$), and a significant correlation between a male's initial, mean, or final size and the number of nights he spent at the pond ($r = -.04, .06, .03$ respectively, $P > > .05$, $N = 617$) was not found.

Male size did change during the study, so to determine the influence of male size on mating success I calculated the probability of mating for males of each size class. I divided the number of matings by males of each size by the total number of frog-nights at each size. For example, if 2 males each spent 10 nights at the breeding site when they were 26 millimeters, and these males obtained a total of 5 matings on these nights, the probability of mating at 26 millimeters would be 0.25 (5 matings/20 frog-nights). The data from 751 matings over 4456 frog-nights demonstrate that a significant positive correlation exists between a male's size and his probability of mating (Spearman rank correlation, $r_s = .71$, $P < .05$, $N = 11$; fig. 4.6).

Table 4.2 Size of Mated Males Relative to the Mean and Median Size of Males Present

	$< \bar{x}$ SVL	$> \bar{x}$ SVL	$= \bar{x}$ SVL	Total
Observed	321	429	1	751
Expected	375	375	—	—
				$X^2 = 15.6$
				$P < 0.005$
	$<$ med SVL	$>$ med SVL	$=$ med SVL	Total
Observed	288	363	100	751
Expected	325.5	325.5	—	—
				$X^2 = 8.2$
				$P < 0.005$

As in the 1978 study, the data on male mating success were analyzed on a nightly basis to test further the hypothesis that females preferentially select larger males. Table 4.2 shows the expected number of matings with males larger and smaller than the mean male size if females chose males randomly with respect to male size. Again, significantly more of the mated males were larger than the mean male size on the night they mated. A statistical assumption of this analysis is that male sizes are distributed normally on each night. A visual inspection of the data suggests that this is true, but on some nights too few males were present to test statistically this assumption. However, a similar analysis can be made without the assumption of a normal distribution by using the median, instead of the mean, male size. The mated males were often the median size, but significantly more of the mated males were larger than the median size than would be expected given random mating (table 4.2).

These results are in accord with those of the pilot study. They show that the more nights a male spends at the breeding site, the more likely he is to acquire mates. However, once a male is at the breeding site, and female choice becomes one of the factors determining male mating success, larger males are more likely to acquire mates. These data, together with the behavioral observations of female mate choice, lead to the conclusion that females preferentially select larger males as mates.

Several possible factors could result in females choosing larger males. One factor already dismissed is the control of higher quality resources by larger males. Another possibility is that females preferentially select males that display in certain areas of the breeding site, even though the females do not utilize resources in the display area. This phenomenon is well known in lekking birds; females tend to mate preferentially with males displaying at a certain position on the lek—usually the center (e.g., Kruijt and Hogan 1967; but Bradbury and Gibson 1983 question the generality of this phenomenon in birds). Why female birds prefer these males is not known. Competition among males for the center of the lek does exist, and some researchers have suggested that females choose males from the center because these will be the most highly competitive males (e.g., Alexander 1974). Others have suggested that the center of the lek is the area safest from predation and that this is the primary selective advantage influencing female choice in lekking birds (see chap. 8).

Regardless of why some female birds choose males from the center of a lek, a position effect was not found at Kodak Pond. Males were not aggregated in clusters but were evenly spaced around the perimeter of the pond (see fig. 3.11). By definition then, a center-edge effect cannot exist because no center or edge exists. If the males had a preferred area, we would expect to find distances between males increasing with distance from the preferred site due to male-male competition for these calling areas. No such distribution was noted. Another observation that tends to argue against a preferred calling site was the behavior of the males. After a male was marked and returned to the pond, he swam along the perimeter, stopped, and called. If he stopped near another male, the intruder was quickly displaced by the previous male. He continued to swim, stop, and call until he reached a calling site from which he was not displaced. The male usually spent the rest of the night calling from that site. All males in the pond tended to remain at the same calling site for the duration of each night, but they did not necessarily return to the same site on successive nights. Although male-male aggressive interactions were common among túngara frogs, these behaviors function in maintaining distances between individuals rather than defending a specific site.

Neither differential access to better territories or calling sites explains the greater mating success of larger males. An obvious alternative is that females are attracted preferentially by some characteristic of the advertisement call, which consequentially results in their selecting larger males. The studies by Whitney and Krebs (1975) and Fellers (1979a, 1979b) reviewed in chapter 1 showed that males calling more frequently, more intensely, or in areas where their calls are better transmitted, are more likely to acquire mates. This result is not surprising and might be a general phenomenon among anurans, even though we do not know if the males calling more intensely or from "better" sites are the same individuals night after night. I was interested to learn whether these sorts of differences in calling behavior would explain any of the size-biased variation in male mating success in túngara frogs. Therefore, I tested the prediction that females are attracted preferentially to larger males because they produce more intense calls or call more frequently.

The censuses of male behavior indicated whether a male was in the inflated or deflated position, and therefore indicated whether or not a male had been calling recently. This point-sample method (Dunbar 1976) provided an estimate of the relative amount of calling by males and allowed me to test whether a male's size influenced his probability of calling. Since males were not present at the breeding site for the same number of census periods, I calculated the proportion of censuses in which a male present at the breeding site was calling (i.e., censuses calling/censuses present). A significant correlation between a male's initial, mean, or final size and his probability of calling ($r = -.05, -.04, -.03$, respectively, $P >> .05, N = 617$) was not found.

The amount of calling was measured in two other ways. In a study of the energetics of calling (see chap. 7), we measured the frog's size (in this case its mass) and his call repetition rate over a 30-minute period. A significant correlation between male size and his call repetition rate ($r_s = -.33$, $P > .05, N = 9$) was not found. As reported in chapter 3, we also counted the total number of calls produced by individual males in an entire night; again, a significant correlation between male size and the number of calls produced was not found ($r_s = -.26, P > .05, N = 5$). These latter two measures are from small sample sizes. However, in each case a trend toward a negative association between male size and the amount of calling does exist. Why this should be is not clear, but one purely speculative suggestion is that for a call to be produced the air in the male's lungs must travel from the lungs across the vocal cords into the vocal sac and back into the lungs. Since a smaller frog will have a shorter distance for the air to traverse (i.e., lungs to vocal sac to lungs), it might produce calls more rapidly. Regardless of the mechanism, and whether or not the trend is biologically meaningful, it is in the direction opposite to that predicted by the hypothesis that females choose larger males as mates because these males call more.

I also measured the call intensity of 5 males. The measurements were made 50 centimeters directly in front of the calling frog. The sound pressure levels of the first 10 calls recorded were averaged, and the male's size was recorded. A significant correlation did not exist between male size and call intensity ($r_s = .005, P >> .05, N = 5$). Although the sample size for the size-intensity relationship also was small, these results agree with those of Passmore (1981), which showed that there need not be an intraspecific relationship between these two factors. (Gerhardt 1975, however, did show that larger *Bufo americanus* produced more intense calls.) Even if larger *P. pustulosus* males did produce more intense calls, this does not necessarily mean that these calls always would be more intense than the calls produced by smaller males when the calls reached the female. Besides the effect of distance between male and female on call intensity, microclimatic events and, more importantly perhaps, differential vegetation structure will greatly influence sound intensity (Wiley and Richards 1978; Michelson 1978). Although female frogs might be attracted by more intense sounds, this would not result in females being attracted preferentially to the calls of larger males.

Summary of Male Mating Success and a Comparison with Other Species

In the pilot study in 1978 and in the more detailed study in 1979, the operational sex ratio was greatly skewed toward males on most nights, predicting intense competition among males for mates. As competition for mates increased, so did the potential for higher variance in male reproductive success, and thus the intensity of sexual selection.

Variance in male reproductive success originates from two sources. The first is variance in the clutch size of females. If all males fertilized the same number of clutches, variance in male reproductive success would exist because some males would have mated with females having larger clutches. The second source of variation in male reproductive success is due to the variance in the number of times a male mates. These two factors define the total intensity of selection on male reproduction. The latter factor defines the intensity of selection due to sexual selection only, but it does not specify the agent of selection (i.e., male competition or female choice).

Wade and Arnold (1980; see also Wade 1979) propose a method for estimating the intensity of sexual selection. Kluge (1981) slightly modified their technique and examined data from a number of anuran species. Following Kluge's notation, I will use the method proposed by Wade and Arnold to estimate the intensity of sexual selection on male reproduction in 1978 and 1979.

The variables needed are the mean and variance of male mating success (x_s and s^2_s, respectively) and female clutch size (x_{fe} and s^2_{fe}, respectively). These data are listed in table 4.3. The male-mating-success variables were reported earlier in this chapter for 1978 and 1979, and the clutch-size data are from a laboratory study of fertilization rates conducted in 1980 (see chap. 6), and are used in estimating the intensity of selection for both years.

Table 4.3 Parameters Estimating Intensity of Sexual Selection for *Physalaemus pustulosus*

		1978	1979
Male mating success			
	$\bar{x}_s$	0.57	1.22
	s^2_s	0.83	3.14
Female clutch size			
	$\bar{x}_s$	234.7	234.7
	s^2_{fe}	9528.6	9528.6
I_m		2.86	2.25
I_s		2.55	2.11

Note: Parameters are defined in text.

Following Kluge, variance in male mating success is:

$$s^2_m = (\bar{x}_s \, s^2_{fe}) + (\bar{x}^2_{fe} \, s^2_s).$$

The intensity of selection on male reproductive success is:
$$I_m = s^2_m/(\bar{x}_s \bar{x}_{fe})^2.$$
And the intensity of sexual selection on male reproduction is:
$$I_s = s^2_s/\bar{x}^2_s.$$
One can see intuitively that I_m measures the intensity of selection due to variance in male mating success as well as clutch size, while I_s only considers the parameters affected by sexual selection—that is, the variance in male mating success. I_s/I_m gives the proportion of selection on male reproduction that is due to sexual selection alone.

The measure of I_s for the 1978 and 1979 study of *P. pustulosus* fall within the range, but at the higher end of the range, for the same measures calculated by Kluge (1981) for a variety of other species of frogs (table 4.4). (Kluge also estimates I_s for *P. pustulosus* from the study in 1978. Our measures differ slightly because the measures for clutch size that I used were from a larger sample size that was not available when Kluge reported his estimates.)

Table 4.4 Sexual Selection Intensities for Some Anuran Species

Species	E or P	CV	I_s	I_s/I_m
Rana sylvatica	E	231.5	4.0	.87
Bufo bufo	E	204.7	5.0	.85
Bufo exsul (1977)	E	214.1	5.2	.90
Bufo exsul (1978)	E	138.5	2.4	.90
Bufo typhonius	E	127.7	—	—
Bufo americanus	E	168.5	—	—
Rana temporaria	E	116.2	1.4	.81
Bufo canorus (1976)	P	286.6	13.6	.96
Bufo canorus (1977)	P	159.9	2.0	.93
Bufo canorus (1978)	P	145.3	1.7	.93
Bufo canorus (1979)	P	131.3	1.8	.94
Hyla versicolor	P	181.5	3.0	.89
Rana clamitans (1974)	P	106.2	—	—
Rana clamitans (1975)	P	143.2	—	—

Table 4.4, *continued*

Species	E or P	CV	I_s	I_s/I_m
Rana catesbeiana (1976)	P	117.6	1.4	.87
Rana catesbeiana (1977)	P	122.5	1.5	.92
Rana catesbeiana (1978)	P	140.4	1.9	.95
Hyla rosenbergi (1977)	P	125.6	1.4	.99
Hyla rosenbergi	P	133.6	1.7	.99
Centrolenella colymbiphyllum	P	104.1	1.1	.99
Centrolenella fleischmanni	P	72.9	0.5	.99
Centrolenella valerioi	P	83.7	0.7	.98
Physalaemus pustulosus (1978)	P	159.8	2.86	.89
Physalaemus pustulosus (1979)	P	145.2	2.25	.94

Source: All the data on mating success, except for *Physalaemus pustulosus*, are from Kluge 1981, tables 13 and 15. The designation of breeding type (P or E) is from Wells 1977a, except for *Physalaemus pustulosus* (personal observations) and the three species of *Centrolenella* (Roy McDiarmid, personal communication).

Note: CV = coefficient of variation of male mating success; I_s = the intensity of sexual selection on male reproduction; I_s/I_m = the proportion of the total selection on male reproduction due to sexual selection for various species of frogs classified as either prolonged (P) or explosive (E) breeders.

The data presented by Kluge allow a comparison of the variance in male mating success and the intensity of sexual selection on males of *P. pustulosus* with that of other species. Wells (1977a) predicted that the variance in male reproductive success is likely to be greater in prolonged breeders than in explosive breeders. Gatz (1981) refuted this hypothesis. He stated that Wells predicted that the mean-to-variance ratio of male mating success in explosive breeders should be significantly less than 1, while the same ratio for prolonged breeders should be significantly greater than 1. For the data in hand, Gatz showed that this notion is not true. Although Wells's prediction suggests differences in the mean-to-variance ratio of mating success between explosive and prolonged breeders, nowhere does Wells make the precise numerical predictions put forth by Gatz.

Although the mean-to-variance ratio used by Gatz to compare prolonged and explosive breeders is an appropriate statistic, I will use the coefficient of variation (100 × standard deviation/mean) to compare different species, since this measure has been used by others for this purpose (e.g., Payne and

Payne 1977; Kluge 1981). No significant difference exists between the coefficients of variation of male mating success for the explosive and the prolonged breeders listed in table 4.4 (Mann Whitney U test, $U = 36$, $P > .10$); and contrary to the prediction by Wells, the median coefficient of variation for explosive breeders, 168.5, is higher than that for the prolonged breeders, 133.6. If we compare the intensity of sexual selection on males, which should be proportional to the coefficient of variation of male mating success (Kluge 1981), the intensity of sexual selection is actually significantly greater for explosive breeders than it is for prolonged breeders (I_s explosive = 5.1, I_s prolonged = 1.7; $U = 14$, $P = .05$).

These results are surprising because the prediction made by Wells is logical, and this prediction also follows from other theoretical treatments of mating systems (Emlen 1976; Emlen and Oring 1977). Gatz seemed to imply that Wells's prediction is falsified because of the importance of female mate choice in explosive breeders, a selection force Wells did not consider to be very important in explosive breeders. Although I agree that female choice does seem to play a more important role in these species than was previously appreciated (see chap. 1; Sullivan 1983), it does not explain the contradiction between data and theory because neither the coefficient of variation nor the intensity of selection distinguishes between male competition and female choice.

The prediction by Wells should hold if the average operational sex ratio each night at the breeding site is lower (i.e., the proportion of males to females) for explosive breeders than for prolonged breeders—regardless of the relative importance of male competition and female choice. This result is expected since in explosive breeders females are receptive over a shorter period of time. However, this assumption need not be true if: (1) the population sex ratio of explosive breeders is more male biased than that of prolonged breeders (perhaps due to relatively earlier maturation of explosively breeding males or higher predation intensities on males of prolonged breeding species); (2) any given male of a prolonged breeding species is only present at the breeding site for a small portion of the breeding season; or (3) the females of species with prolonged breeding seasons produce more than one clutch and have a short time interval between clutchs. My guess is that the discrepancy exists between the prediction by Wells and the available data because more male-biased operational sex ratios tend to occur in explosively breeding frogs than in prolonged breeders, possibly due to all of the three factors just listed (see also Kluge 1981). Clearly, we know much more about anuran mating systems now than when Wells wrote his stimulating review, but we need to know much more before we can hope to understand the diversity of anuran reproductive behaviors observed in nature.

The selection estimates for túngara frogs and for other frogs are lower than those generally reported for birds. Kluge suggests that measures of variance of reproductive success might be inflated in avian studies because behaviors that do not lead to reproduction, such as mountings and some copulations, often are scored as successful fertilizations. This potential problem always arises in studies of species with internal fertilization, and the problem should not be ignored. However, there might be a biologically more significant factor contributing to the different variances among taxa. Male frogs rarely mate more than once a night due to the prolonged duration of amplexus. This constraint of "servicing time" (Emlen and Oring 1977) decreases the ability of a few males to monopolize most of the matings, and thus, decreases the potential variance in male mating success. This time constraint is not present for many internal fertilizers, where a large number of females can be fertilized in rapid succession.

The data from túngara frogs show that skewed mating success among males is due to sexual selection, and that the intensity of sexual selection is greater than that of about two-thirds of the frog species for which data are available. The next question is, which males are more likely to acquire mates? The best predictor of a male's mating success is the number of nights he spends at the breeding site. Why some males spend more time breeding than others is not known; energetic constraints seem to be important (chap. 7), but clearly a male's size does not influence this aspect of his breeding behavior.

Data from both years indicate that size influences a male's mating success once he is at the pond, where sexual selection can exert its influence. Whether the data are combined for all nights or, more importantly, if they are analyzed on a nightly basis, larger males are more likely to acquire mates. Since the behavioral observations indicate that female mate choice should influence male mating success, one must conclude that females choose larger males. Therefore, size-based differences among males should influence female choice in such a way that females are attracted preferentially to larger males.

Several factors could potentially influence mate choice by females. The possibility that larger males defend better resources is immediately dismissed. The male's spatial location also is not important, based on my observations of male and female behaviors.

The advertisement call certainly plays some role in mate choice, if only at the species level. I proposed the hypothesis that female choice of larger males might result from preferential phonotaxis by females, based on differences in quantitative or qualitative aspects of the advertisement call. The first hypothesis was rejected. In the next chapter, I discuss the various aspects of how the advertisement call functions in male reproductive behavior, and I test the second hypothesis—that female choice of larger males results from qualitative differences in the advertisement call.

5 The Function of the Advertisement Call

The life of the túngara frog appears to be dominated by one behavior pattern—calling—partly because the frogs are rarely seen when they are not at the breeding site. However, as we shall see, when túngara frogs are at the breeding site, calling consumes a fair amount of their time and energy and is quite a dangerous venture. But in this species, a male attempting to reproduce seems to have no other viable options besides advertising his presence until a female chooses him as a mate. As with most other frogs, the advertisement call is the dominant feature of the *P. pustulosus* mating system. In this chapter, I examine how variation in the advertisement call might influence differential mating success among males; this first entails a detailed analysis of the function and structure of the advertisement call.

A Complex Advertisement Call: Description of Call Components

The importance of the anuran advertisement call in species recognition has been investigated thoroughly (see chap. 1). This role of the call has probably resulted in much of the call's stereotypy within a species, and has also led many researchers to concentrate on stereotypy in frog advertisement calls. However, the advertisement call of *P. pustulosus* is interesting because of its variability. This species is unusual in that individuals produce calls of varying complexity, although this type of call structure is now being discovered in other tropical frogs. Rand first investigated the túngara frog's advertisement call in the early 1960s. He described the varying complexity of the call and demonstrated how this complexity is mediated through male and female behaviors. When first hearing this call, one cannot clearly distinguish that only a single species of frog is producing these sounds: only a closer look and listen reveal that the whines and the chucks come from the same individuals. An initial surprise is that not all individuals produce the same calls. All produce whines, but some produce calls with only whines, some with 1 chuck at the end, and a few produce calls with up to 6 chucks. These call differences are not due to a behavioral polymorphism, that is, the type of call is not specific to the individual; listening for a longer period reveals that each individual is capable of producing all of the call variants.

A detailed analysis of the advertisement call was necessary before I could hope to understand many aspects of the call's function and evolution. I employed three techniques commonly used in bioacoustics to quantify call characteristics. The first is provided by the oscilliscope. This machine produces a graphical representation showing how call energy, on the vertical axis, changes as a function of time, on the horizontal axis (e.g., fig. 5.1). The second device is a sonogram, and it produces a graph demonstrating the change in frequency, on the vertical axis, as a function of time, on the horizontal axis (e.g., fig. 5.2). The power spectrum combines aspects of both the oscillograph and the sonograph (e.g., fig. 5.3). For a fixed duration of time, the power spectrum shows the distribution of energy among frequencies. Since the time period over which the signal is analyzed is often of shorter duration than the entire call, that part of the call that is analyzed must be specified. Together, these three techniques allow a quantitative description of all the physical characteristics of the call.

The temporal features of *P. pustulosus* advertisement calls of varying complexity are shown in the oscillograms of figure 5.1. In a representative call, the whine is 400 milliseconds in duration, regardless of whether or not it is accompanied by a chuck. The whine reaches its peak amplitude in 26 milliseconds, after which the amplitude slowly and steadily declines. In a complex call, the amplitude of the whine is at its minimum immediately before the chuck is produced. The chuck is 42 milliseconds in duration, much shorter than the whine, and is pulsed. The chuck reaches its maximum amplitude in 15 milliseconds, and remains at this level for 10 milliseconds before it begins to decay rapidly. Both figures 5.1 and 5.2 clearly show that the whine and the chuck overlap in time. This finding suggests that the frog must produce the two call components synchronously, as opposed to the whine being produced when air is exhaled and the chuck when air is inhaled, as stated by Altig (1979).

The whine contains extreme frequency modulation; the dominant frequency (the frequency with the most energy) of this call component is approximately 900 Hz, and during the 400 milliseconds of the call it sweeps to approximately 400 Hz (fig. 5.2). Both the whine and the chuck have harmonics, that is, they have frequencies that are multiple integers of a fundamental frequency. Upper harmonics of the whine are sometimes visible in the sonogram, but the 900 Hz to 400 Hz frequency sweep is always the dominant. In a complex call the whine appears to grade into the chuck, the final frequency of the whine being the second harmonic of the fundamental frequency of the chuck. The chuck, in contrast to the whine, has a rich harmonic structure, and the upper harmonics usually contain more energy than the lower ones (fig. 5.2; see also fig. 5.6).

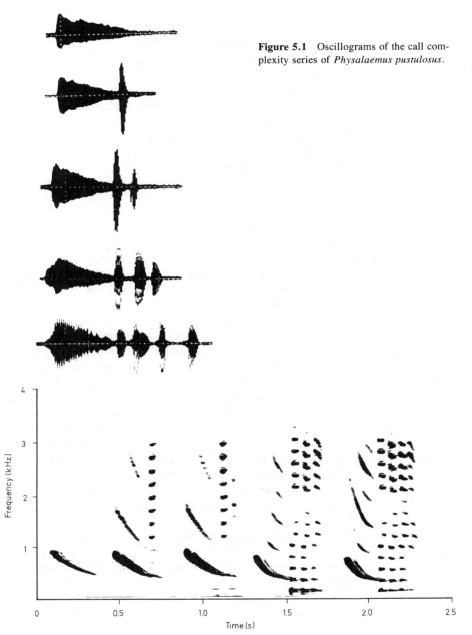

Figure 5.1 Oscillograms of the call complexity series of *Physalaemus pustulosus*.

Figure 5.2 Sonograms of the call complexity series of *Physalaemus pustulosus* (from Ryan 1983a).

For a more detailed analysis of the call, I determined the power spectrum of various parts of a typical whine-plus-chuck call, which is shown in figure 5.3. The analysis of the first 80 milliseconds of this whine (fig. 5.4A) shows that peak energy concentration is at 787 Hz (fig. 5.4B). As determined from

Advertisement Call Function 69

the dominant frequency band in the power spectrum, during this 80 millisecond period the whine has swept from 925 Hz to 700 Hz. The fact that the peak concentration during this time is at 787 Hz is consistent with the oscillogram, which shows that the amplitude of the whine reaches its peak after 26 milliseconds (fig. 5.4A) when the dominant frequency sweep is at

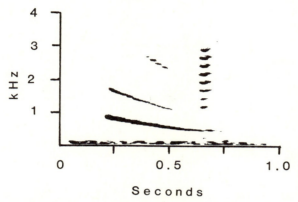

Figure 5.3 Sonogram of the whine-plus-chuck call used for the more detailed spectral and temporal call analyses in figures 5.4 to 5.6.

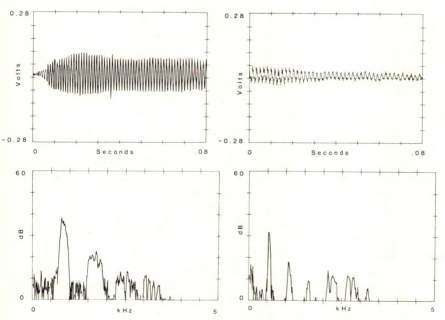

Figure 5.4 A, oscillogram of the first 80 milliseconds of the whine; B, power spectrum of the first 80 milliseconds of the whine.

Figure 5.5 A, oscillogram of the final (i.e., just prior to the chuck) 80 milliseconds of the whine; B, power spectrum of the final 80 of milliseconds the whine.

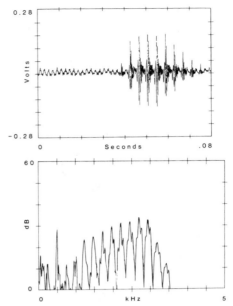

Figure 5.6 *A*, oscillogram of the chuck; *B*, power spectrum of the chuck.

787 Hz (fig. 5.3). Also, a substantial concentration of energy in the second harmonic of the dominant energy band occurs from 1400 to 1850 Hz. Some energy is seen in the third (2100 Hz to 2775 Hz) and fourth (2800 Hz to 3700 Hz) harmonics of the dominant energy band, but clearly no energy is contained in the fifth harmonic (3500 Hz to 4625 Hz; fig. 5.4*B*). The spectral analysis supports the impression from the sonogram (fig. 5.3) that the fundamental frequency of the whine is the dominant frequency, and that in the beginning of the call substantial energy is contained in the second harmonic with less energy in the third and fourth harmonics.

During the final 80 milliseconds of the whine, just before it grades into the chuck (fig. 5.5*A*), the fundamental frequency is still dominant (fig. 5.5*B*). Peak energy concentration is at 550 Hz. The dominant frequency band during this part of the call is narrower than the initial 80 milliseconds because the sweep covers a narrower frequency range during the end of the whine (fig. 5.3). There are discrete peaks of energy at the second (1100 Hz) and third (1650 Hz) harmonics with lesser amounts of energy at the fourth (2200 Hz), fifth (2750 Hz), and sixth (3300 Hz) harmonics (fig. 5.5*B*). The whine, therefore, appears to be produced by a complex vibration pattern in which the fundamental frequency, which changes through time, is the dominant frequency with substantial energy in the third, fourth, and sometimes the fifth and sixth harmonics (fig. 5.5*B*).

The spectral structure of the chuck is extremely different from that of the whine. The power spectrum of the entire chuck (fig. 5.6*A*) shows that the sound is produced by a vibration that has a fundamental frequency, in this call, of 250 Hz (fig. 5.6B). Energy is present in the fundamental frequency and in each of its first twelve harmonics. In some calls, the fundamental frequency and the third, fourth, and fourteenth harmonics are not visible on a sonogram. The power spectrum shows that the second harmonic is relatively intense, perhaps because this frequency also has energy contributed to it by the whine (fig. 5.6B). The relative amplitude of the harmonics decrease from the second to the third harmonic, and then the relative amplitudes tend to increase from the third harmonic until the peak amplitude is reached in the eleventh harmonic. After the twelfth harmonic, the relative amplitude drastically decreases and there is no energy in the fifteenth harmonic. If we compare the amplitude of the chuck with that of the whine, clearly the chuck usually has greater peak amplitude (fig. 5.1), although a fair amount of variation exists in the relative amplitudes of the call components within an individual frog. A comparison of the two power spectra of the whine (figs. 5.4B, 5.5B) with that of the chuck (fig. 5.6B) reveals that the greater relative amplitude of the chuck is due primarily to its rich harmonic structure. The dominant frequencies of each of the whine sections have a greater relative amplitude than the dominant frequency of the chuck.

Mechanism of Vocalization

The advertisement call of *P. pustulosus* has two interesting structural characteristics. The first is the extreme frequency modulation of the whine. Although this call type is common among the Leptodactylinae (Straughn and Heyer 1976), the subfamily that includes *Physalaemus*, very few North American frogs have calls with such extreme frequency sweeps. Another trait of the call, and one that caused some initial confusion (e.g., Altig 1979), is that the whine and the chuck temporally overlap. How a frog could produce a call that had two distinct components overlapping in time was not clear.

Before Martin presented his analysis of the mechanism of *Bufo* vocalizations in 1972, the understanding of how frogs made calls, and specifically how they regulated the amplitude and frequency of the call, was rudimentary. At this time, the theory of the vocal mechanism of the complex songs of birds was much more advanced. For hundreds of years, scientists and naturalists speculated as to how birds might produce their complex songs. In 1968, Crawford Greenwalt presented his model in a book that has now become a classic in animal communication: *Bird Song: Acoustics and Physiology*. According to this model, the vocal apparatus of birds—the syrinx—contains two independent sound sources, one in each bronchus, which enables the bird to produce two "phrases" simultaneously. (Brackenbury 1982 reviews the current understanding of the process.) Aware of these results and struck by the temporal overlap of the túngara frog's whine and chuck, Rand sent túngara frog and treefrog call recordings to Greenwalt and asked his opinion regarding the possible mechanism of vocalization. In a letter Greenwalt replied:

> It looks as if the acoustical system of tree frogs is quite similar to that of birds; that is to say, there must be a membrane vibrating in a constricted passage with musculature somehow attached to permit modulation in both frequency and amplitude. The frequency range and the frequency of modulation indicate that the acoustical system must have a very low mass and hence is similar to birds in that characteristic also.
>
> There is no very good evidence that frogs have the two acoustical systems which birds possess. There is a hint in *Engystomops* (= *Physalaemus*) *pustulosus* that something of this sort might exist, but frankly I doubt it.
>
> Of course, you should also do some anatomical investigations to see whether you can actually find the features I've described and get some idea as to where they are and how they might operate. (Rand 1976)

Together with George Drewry and W. Ronald Heyer, Rand did just that and found very much what Greenwalt had predicted (Drewry, Heyer, and Rand 1982). Like calls of most anurans, the túngara frog call is produced by vibrations of the vocal cords, which result when air is forced from the lungs through the larynx. The frequency modulation of the whine is produced by a set of muscles that, when constricted, change the shape of the larynx in such a way that the tension of the vocal cords is increased. As air passes from the lungs over the vocal cords and into the vocal sac, the muscles relax, decreasing the tension of the vocal cords. The frequency at which any membrane vibrates is determined primarily by two factors: its mass (a product of its length and mass per unit length) and tension (Roederer 1975). When the muscles associated with the larynx decrease the tension of the vocal cords, this action consequently decreases the frequency at which the vocal cords vibrate, thus producing the whine. A similar mechanism probably is responsible for the production of frequency modulated calls in other frogs (Martin 1972).

Drewry, Heyer, and Rand (1982) suggested that a further change in larynx shape, also caused by associated laryngeal muscles, brings a set of two fibrous masses in the larynx into the air flow. The fibrous masses are coupled both acoustically and physically to the vocal cords. As the fibrous masses vibrate, they cause the vocal cords to vibrate at twice the fundamental frequency of the vibration of the fibrous masses, which is why the final frequency or the whine in a complex call is usually the same as the second harmonic of the chuck. It is this vibration that produces the chuck, and the existence of two partially coupled vibrating structures (vocal cords and fibrous masses) results in the two call components overlapping in time.

The advertisement calls of other species of *Physalaemus* are simple (i.e., only one call component) and resemble either the *P. pustulosus* whine or chuck (Barrio 1965; Drewry, Heyer, and Rand 1982). Túngara frogs are the only members of this genus that have evolved a multi-component call.

Social Mediation of Call Complexity

A male túngara frog calling alone in a pond usually produces only the whine component of the call, which constitutes the simple call. As chorus size increases, so does the complexity of the calls that males produce—that is, they produce calls with more chucks. This finding suggests that call complexity is mediated through social interactions among males; it was one of the first questions Rand addressed when he began to investigate the complex call of *P. pustulosus*.

Table 5.1 Vocal Responses of Male *Physalaemus pustulosus* to Various Stimuli Presentations

	Whine Only	Whine plus 1 Chuck	Whine plus 2 Chucks
With stimulus	15	239	46
Without stimulus	276	24	0
$X^2 = 455.8; P < 0.005$			
Stimulus Presented	*> Complexity*	*< Complexity*	*No Change*
Loud	20	0	2
Soft	0	20	2
$X^2 = 40.0; P < 0.005$			
Fast	15	2	3
Slow	0	18	2
$X^2 = 28.0; P < 0.005$			

Source: Rand and Ryan 1981.

Rand recorded the calling behavior of a male in the absence of other males. He then played the male a stimulus—a whine-only call—and simultaneously recorded the vocalizations of the calling male. The results of this experiment are presented in table 5.1. In 18 no stimulus–stimulus alternations the male produced 600 calls. A chi-square analysis showed that the complexity of a male's response increased in the presence of the stimulus—that is, the male produced more chucks when he heard the stimulus call, which he apparently perceived as another calling male. Not only did the presence of another calling male influence call complexity, but the distance of the nearest male and the number of calling males were also important. Rand demonstrated this by presenting males with stimulus calls that were more or less intense, the assumption being that more intense calls would be perceived by the experimental male as originating from a closer neighbor than the less intense call (the farther away a male, the less intense his call). He then presented 10 males, one at a time, with soft-loud-soft-loud-soft stimulus alternations and recorded their responses. He determined whether the mean complexity (i.e., the mean number of chucks per call) increased or decreased between stimulus types. As table 5.1 shows, males always decreased mean call complexity when the stimulus shifted from a loud call to a soft call, and call complexity increased when the stimulus shifted from soft to loud. Therefore, males increase the number of chucks in their calls when they are closer to other calling males.

Rand also experimentally varied the apparent number of calling males. When males call they usually do so at a maximum rate of about 1 call every 2 seconds. Therefore, a call played at a rate of 1 call every second should be perceived by the experimental male as the calling of two neighboring males. Rand presented 11 frogs with slow-fast-slow-fast-slow call rates and, as above, he recorded the change in mean call complexity of the male between stimulus types. Again, social factors influenced call complexity; the males increased the number of chucks in their calls in response to the calling of other males.

Not only do túngara frogs produce calls that vary in complexity, but the social milieu in which the calls are produced influences call complexity. These early experiments of Rand did much to dispel the view of frogs as automatons sitting in murky swamps producing the identical signal again and again and again, oblivious to their social environment.

Why do males add chucks to their calls? Since one of the commonly assumed functions of the advertisement call is to attract females (the other is male spacing), Rand tested the differential ability of simple and complex calls to elicit phonotaxis (attraction to a particular sound) from females. A female was placed in a test arena, where she sat in between two speakers broadcasting túngara frog calls. Rand first demonstrated that both simple and complex calls were capable of eliciting phonotaxis from females when the calls were broadcast in isolation. He then presented females with a whine from one speaker and a whine plus chuck from the other speaker, and he recorded the response of the female. These experiments showed that females were attracted preferentially to the complex call (table 5.2). This finding suggests that males increase their call complexity in the presence of other males in order to increase their ability to attract mates.

In these experiments, Rand broadcast the whine and the whine plus chuck at an intensity such that the maximum amplitudes of the calls at the site of the female were equal. However, because the whine and the whine plus chuck have different structures (fig. 5.1), when the maximum amplitudes of the calls are equal, the energy content of the calls are different. Because of this "discrepancy" some researchers have criticized the female-choice experiments just described, suggesting that the female preferentially responds to the whine plus chuck simply because it contains more total energy than the

Table 5.2 Female *Physalaemus pustulosus* Response to Simple versus Complex calls in a Paired-choice Experiment

Experiment	Whine	Whine plus Chuck	P
Rand	1	14	.001
Ryan	1[a]	8[b]	.039

[a] Intensity of Whine = 75 dB SPL.
[b] Intensity of Whine plus chuck = 70 dB SPL

whine. However, these criticisms confuse ultimate and proximate causation, an unfortunate but common occurrence in biology (Mayr 1982). Regardless of the complex call characteristics that are responsible for preferential phonotaxis by females, the experiments clearly demonstrate that males producing complex calls are more likely to attract females than are males producing only simple calls. In the ultimate, evolutionary sense, the precise sensory mechanism that results in the female's preferential response is not of crucial importance. Nevertheless, I duplicated the female-choice experiments by Rand, except that I presented the female with a whine at an intensity of 75 dB SPL and a whine plus chuck at an intensity of 70 dB SPL. Females still preferred the more complex call even though it was broadcast at a lower maximum amplitude than the simple call (table 5.2).

The Paradox of the Call Complexity Series

The above results represent a paradox: If males vocalize in order to attract mates, and females are attracted preferentially to complex calls, then why do males not increase their ability to attract mates by always producing the calls more attractive to females? The evolution of behavior can be considered in terms of its potential costs and benefits, and by keeping in mind that constraints on the evolution of behavior exist—that is, by asking how the behavior might increase or decrease an individual's fitness, and what factors might constrain the behavior from evolving in response to selection. The obvious benefit to producing complex calls is that the male is more likely to mate, thus increasing the proportion of his alleles in future generations. But the fact that males do not always produce complex calls suggests that there must also be some disadvantage in producing chucks.

One possible "cost" is energy. If it were energetically more demanding to produce complex calls, then a male calling alone with no competition from other males could save energy by producing simple calls and still not decrease the possibility that he would attract a female. The amount of energy expended for sexual advertisement has been suggested as a partial explanation for the adaptive significance of some noncalling reproductive behaviors by frogs (e.g., Perrill, Gerhardt, and Daniels 1982; Gerhardt 1983). However, at the time I was conducting this study no measure existed of energy expended during a sexual display by any vertebrate. While I was on

76   Chapter Five

Barro Colorado Island, George Bartholomew and Terry Bucher of U.C.L.A. were studying the physiological ecology of birds, moths, and beetles, and their studies included measures of the rate of oxygen consumption of these animals. Rand had predicted a cost associated with complex calls, and several years earlier he had suggested to Bartholomew that energy expenditure during calling by túngara frogs should be measured. Unfortunately, Rand was not available to join us at the time we conducted the initial measurements of energy expended during calling, but he did join us in later studies of túngara frog reproductive energetics.

We investigated several aspects of the reproductive energetics of *P. pustulosus*, and these results are presented in chapter 7. Here, however, I will consider those data we secured to test the hypothesis that producing chucks increases the amount of energy expended for sexual advertisement. In 1980, we determined the amount of aerobic energy expended during calling by measuring the amount of oxygen consumed by calling males. The amount of oxygen consumed during a behavior is indicative of the amount of energy expended to perform that behavior. We collected male frogs on Barro Colorado Island and placed them in glass respirometers (fig. 5.7). The males were placed on a laboratory bench on the ground floor of Kodak House. The laboratory walls were screen and, therefore, the frogs were within earshot of the túngara frog chorus behind Kodak House. A speaker near the respirometer broadcasted a whine plus chucks continuously during the approximately 30 minutes of the experiment. As the calls were played to the male in the respirometer, we recorded his vocalizations with another tape recorder. After the experiment, we determined the amount of oxygen consumed by the

Figure 5.7   A male túngara frog in a respirometer.

calling male by comparing the amount of oxygen in the experimental respirometer to that in an identical respirometer without a male. The difference in the amount of oxygen between the two respirometers was due to the oxygen consumed by the calling male, allowing us to calculate the male's rate of oxygen consumption. With the tape recording of the male, we determined the number of whines and chucks produced by the male during the period his oxygen consumption was measured. Therefore, we could determine the amount of oxygen consumed for each call (= per whine) and for each chuck.

We measured the oxygen consumption for 9 frogs while they were calling, and we determined how the number of whines, the number of chucks, and the male's mass influenced his rate of oxygen consumption. The mean rate of oxygen consumption for a calling male was 1.13 milliliters of oxygen/hour (standard deviation = 0.13), and these males produced an average of 774 calls (standard deviation = 463) with 534 chucks (standard deviation = 607) per hour. The average amount of oxygen consumed, above that needed for maintenance per call, was 0.0015 milliliters. The results of a multiple regression analysis indicate that the number of whines a male produced explains most of the variation in oxygen consumption among calling males, but that the number of chucks does not explain a significant amount of the variation (tables 7.2, 7.3; see chap. 7 for a complete interpretation of *P. pustulosus* reproductive energetics). Therefore, no evidence was found to indicate an increased energetic cost associated with increased call complexity. The hypothesis that túngara frogs vary the number of chucks to conserve energy is rejected.

Predation is another cost which often has been implicated in the evolution of animal communication (e.g., Marler 1955; Moynihan 1970; Smith 1977). Marler (1955) suggested that predation has selected for relationships between the structure and function of different types of bird calls. Avian alarm calls presumably function as warning to others of the presence of predators, and one assumes that there should be selection for a call not easily located by the predator. Mobbing calls, on the other hand, serve to elicit the participation of others in the mobbing of predators. An animal that produced a mobbing call presumably would attempt to advertise its location accurately. A survey of mobbing calls of different species reveals that typically these calls are short interrupted sounds having a wide range of frequencies. These traits should increase the localizability of the sound to a receiver using binaural cues, such as differences in time of arrival, intensity, or spectral structure for localization of sound (e.g., Knudsen 1980). On the other hand, the structure of alarm calls probably renders these calls more difficult to locate for animals using binaural cues (e.g., Knudsen and Konishi 1979). C. Brown (1982) offers experimental support of Marler's hypothesis, and suggests that this hypothesis should be true whether the mechanism underlying binaural comparisons is a pressure receptor, as in mammals (e.g., Heffner and Heffner 1980), or a pressure gradient receptor, as has been suggested recently for some birds (Lewis and Coles 1980). Brown also points out some experimental problems with a previous study (Shalter 1978) that rejected Marler's hypothesis.

Sonograms of avian mobbing calls and alarm calls are shown in figure 5.8. When one compares the avian alarm calls with the whine call component of the túngara frog call, a striking resemblance in call structure is apparent (figs. 5.2, 5.8). The mobbing calls and the chucks also share similar structural characteristics. As predation presumably has had some role in the evolution of bird song structure, Rand proposed an analogous role for predation in the evolution of the call complexity series of the túngara frog. Specifically, he predicted that males only add chucks to their calls in response to vocal competition from other males because producing chucks increases the predation risk of the male. This hypothesis was interesting, but at the time Rand made this suggestion (1965), no predators were known to orient to frog calls.

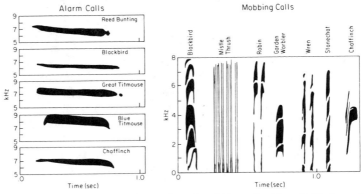

Figure 5.8 Sonograms of avian alarm calls and mobbing calls. From P. Marler and W. J. Hamilton III, *Mechanisms of animal behavior*. Copyright © 1966 by John Wiley and Sons, Inc. Reprinted by permission of John Wiley and Sons, Inc.

While on Barro Colorado Island in 1979, I received a manuscript from Merlin Tuttle of the Milwaukee Public Museum. Tuttle studied bats, and he knew I was studying frogs on Barro Colorado Island. He wanted my comments on some preliminary results he had collected concerning bats that eat frogs. His results were preliminary—and they were startling! Tuttle had documented that on Barro Colorado Island a species of bat, *Trachops cirrhosus* (Phyllostomatidae), eats frogs, and very likely the bats used frogs' vocalizations to locate their prey (fig. 5.9).

As Tuttle had realized, this finding could have important implications for the evolution of frog vocalizations. I suggested that we jointly investigate the possibility that túngara frogs are more likely to be located by predators when they produce chucks. (*Trachops* seemed to be the perfect predator for such an experiment.) This was the first problem we addressed when Tuttle arrived on the island, and we have been working together on bat-frog interactions ever since.

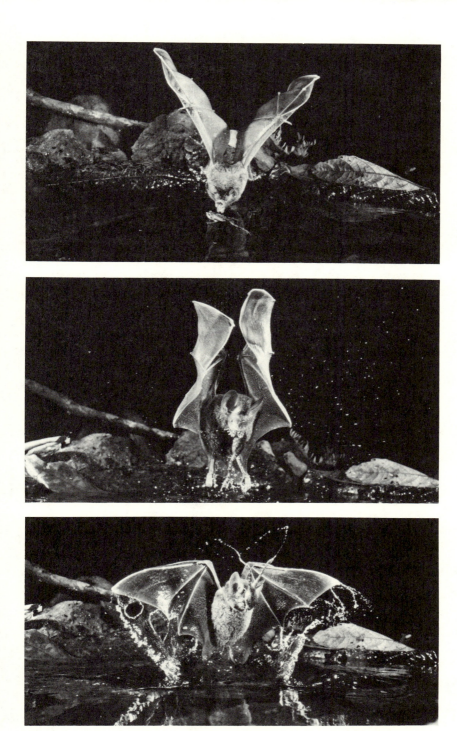

Figure 5.9 *Trachops cirrhosus* capturing a male *Physalaemus pustulosus* (photo by Merlin D. Tuttle).

In chapter 8 I will describe the numerous ways in which bat predation seems to have influenced the reproductive behavior of *P. pustulosus*. But now I will only consider those data that address the predation risks of males producing complex calls. We tested the preferential response of *T. cirrhosus* to simple and complex (whine plus 3 chucks) túngara frog calls. The first night we captured a bat in a mist net and placed it in a flight cage. We then played a túngara frog call from a speaker. In an instant the bat flew from its perch and made a beeline directly to the speaker, actually landing on it (fig. 5.10). The bat jumped to the ground and pulled itself around the perimeter of the speaker, behaving as if there were a frog inside. Obviously we would be able to test bat preference much as I had tested female frog preference for túngara frog calls.

The flight cage served as a test arena for our choice experiments with bats. We placed speakers in two corners, one on each side of the bat (fig. 5.11); one speaker broadcasted a simple call, while the other played a complex call. We sat in the far corner opposite the bat, where we controlled the tape recorders that drove the speakers. We played the calls simultaneously and recorded the bat's response. We used a similar procedure in the field. *Trachops* are attracted to frog calls broadcast from speakers in the field (fig. 5.12), and we have taken advantage of this to lure bats into mist nets, thus increasing our netting success (fig. 5.13). At areas where we thought *Trachops* might be active, we placed two small cassette recorders 4 meters apart. The speaker from one recorder broadcasted a simple call and the other a complex call. Using a night vision scope, we were able to count the number of passes made by the bats over each speaker (fig. 5.14).

Table 5.3 Distances within Which *Trachops cirrhosus* Approached Speakers during Flight Cage Experiments

	Landed on Speaker	Approached (within inches)			
		$0 < 4$	$4 < 8$	$8 < 12$	> 12
Responses	14	15	11	1	3

Source: Ryan, Tuttle, and Rand 1982.

We tested the responses of 5 bats to calls in the flight cage. To prevent the bats from habituating to the calls, we turned off the speakers immediately after a bat left its perch. Although a response was noted if the bat flew within 1 meter of the speaker, even in the absence of a continuous acoustic cue the bats usually approached much closer than 1 meter (table 5.3). From our previous experience, we knew that if a speaker continued to broadcast frog calls, the bat would soon land on the speaker. Each of the 5 bats were attracted more often to the complex call than to the simple call. In 3 of the 5 cases, these preferences were statistically significant; and when the data are combined for all trials, an overall preference for the complex call can be seen ($X^2 = 40.1$, $P < 0.005$; table 5.4).

Figure 5.10 A *Trachops cirrhosus* flying towards a speaker broadcasting frog calls in the flight cage (photo by Merlin D. Tuttle).

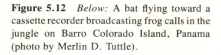

Figure 5.11 *Left:* A figure of the flight cage, showing the positions of the speakers, the observers, and the bat during tests of bat preference for túngara frog calls.

Figure 5.12 *Below:* A bat flying toward a cassette recorder broadcasting frog calls in the jungle on Barro Colorado Island, Panama (photo by Merlin D. Tuttle).

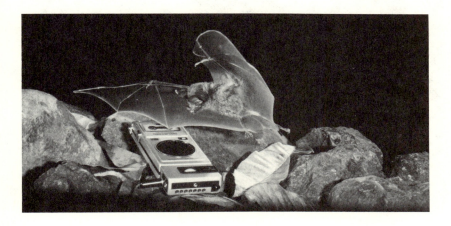

82 Chapter Five

Figure 5.13 The author, Cindy Taft, and Merlin Tuttle (*l-r*) preparing to band a recently netted *Trachops cirrhosus* (photo by Merlin D. Tuttle).

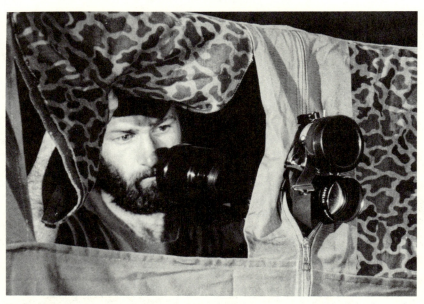

Figure 5.14 The author in the blind with the night vision scope used to conduct observations of the bats' approaches to cassette recorders broadcasting frog calls in the jungle on Barro Colorado Island, Panama (photo by Merlin D. Tuttle).

We conducted the field experiments on five different nights at five different sites on Barro Colorado Island. The results paralleled those from the flight cage (table 5.4). At all but one site, significantly more passes were made by *Trachops* over the speaker producing the complex call compared to the speaker producing the simple call. At the site where no significant difference in the number of passes was found, more passes were made over the speaker broadcasting the complex call. When the data are combined for all sites, an overall preference for complex calls is noted ($X^2 = 51.2$, $P < 0.005$; table 5.4).

Table 5.4 Responses of *Trachops cirrhosus* to Simple and Complex *Physalaemus pustulosus* Calls in the Flight Case (A) and in the Field (B).

A.

Bat	Complex	Simple	P
f 7	8	0	.0039
m 9	4	2	.3437
m12	9	3	.0730
f13	8	0	.0039
f14	10	0	.0001

B.

Date	Site	Complex	Simple	P
2/11/80	Nemesia	56	37	.0307
2/13/80	Barbour	15	8	.1050
2/14/80	Shannon	52	22	.0003
2/15/80	Standley	37	13	.0004
2/17/80	Chapman	8	1	.0195

Source: Ryan, Tuttle, and Rand 1982.

Why the bats prefer complex calls is not clear. These calls might be easier to locate because the structure of the chuck presents cues important for binaural comparisons (Marler 1955; Knudsen and Konishi 1979; Knudsen 1980). These calls also contain higher frequencies that, at first thought, might be more important for localization by a bat, since these animals are well known for their ability to perceive high frequency sounds (e.g., Griffin 1958). However, we now know that *T. cirrhosus* is very capable of hearing the relatively low (for a bat) frequencies contained in the whine (Ryan and Tuttle 1983; Ryan, Tuttle, and Barclay 1983).

Regardless of why the bats are attracted preferentially to these calls, male túngara frogs appear to have a higher risk of bat predation when they produce complex calls instead of simple calls. Certain other predators, such as philander opossums, which also use acoustic cues to hunt túngara frogs, are probably attracted to these calls as well (Tuttle, Taft, and Ryan 1982; chap. 7). These results support the prediction made by Rand 15 years before our experiments, and suggest that the *P. pustulosus* call complexity series has evolved because "it allows a male to adjust call complexity to effect a compromise between maximizing mate attraction ability and minimizing predation risk" (Rand and Ryan 1981).

The Chuck and Sexual Selection by Female Choice

Because females are attracted to calls without chucks, this call component does not seem to be necessary for species recognition. The chucks do increase the ability of a male to attract females, and chucks seem to serve no other role than a behavioral means for males to compete for the attention of females. That the chuck has not evolved under the influence of sexual selection, specifically female choice, would be difficult to deny. And female choice in this species must be a powerful selective force: it has resulted in males changing their calls in a way that is maladaptive in the context of natural selection, because it increases predation risk.

One question left unanswered in the above section is: Why are females attracted preferentially to complex calls? One possibility is that perhaps, like the bats, the female frogs are better able to locate complex calls. Sound localization is still not perfectly understood for birds, for which most of the data exist (e.g., see review in Knudsen 1980). In general, birds appear to use binaural cues; specifically, they seem to compare differences in intensity, time of arrival, phase angle, and spectral structure of a sound at both ears in order to localize sound along a horizontal axis. Monaural cues seem to be important in elucidating the elevation of the sound (Knudsen 1980). But sound locatability by binaural comparisons might present a difficulty for frogs. When using binaural cues, the accuracy of sound localization should increase with the amount of the difference in the cues between each ear. Differences in any of the above cues increase with the distance between the ears. Because the interaural distance of frogs is minute, especially in comparison to some of the wavelengths of the sounds they need to localize, much debate about exactly how frogs locate sound still exists (e.g., Gerhardt and Rheinlaender 1980; Wilczynski, Resler, and Capranica 1981; Gerhardt 1983). Therefore, because a chuck might make a call easier to locate for one kind of animal (a bat), it does not necessarily follow that female frogs also will be able to locate complex calls better than simple calls. However, calls with chucks are slightly longer in duration, contain more total energy, and cover a broader frequency range than calls without chucks. Chucks also have rapid rise and decay times (e.g., fig. 5.6A). Regardless of the mechanism of sound localization by anurans, these characteristics of calls with chucks could make them easier to locate.

Some data address the ability of females to locate calls. Rand presented females with either a whine or a whine plus chuck in the test arena. The floor

Advertisement Call Function 85

of the test arena was marked into 8 equal pie-shaped sections (fig. 5.15). Rand recorded the section the female was in when she departed from the center of the arena (i.e., the inner circle in fig. 5.15). These results are presented in figure 5.15. I compared the number of frogs that departed from the inner circle of the arena in the 2 sections on either side of the speaker to the number of frogs that departed from the other 6 sections for responses to both simple and complex calls. This analysis demonstrates that females were not able to locate the complex calls more accurately than the simple calls (table 5.5), which is also true if the locations where the females left the arena (i.e., the outer circle in fig. 5.15; table 5.5.) are analyzed. These results do not support the hypothesis that females are attracted preferentially to calls with chucks because they are easier to locate.

Table 5.5 Orientation of Female *Physalaemus pustulosus* to Simple and Complex Advertistment Calls

Female's Position	*Whine*	*Whine plus chuck*	*Total*
Inside circle			
Oriented	18	24	42
Not oriented	27	21	48
$X^2 = 1.42; P > 0.10$			
Outside circle			
Oriented	29	26	55
Not Oriented	16	10	26
$X^2 = 0.55; P > 0.10$			

Source: See fig. 5.13.
Note: Females were oriented if they passed over the inner circle or the outer circle of the arena in one of the two sections on either side of the speaker. They were not oriented if they were in one of the six other sections.

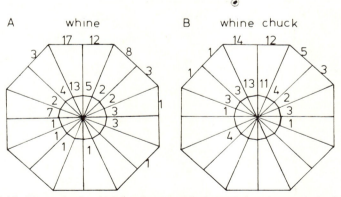

Figure 5.15 The section in which females passed the inner circle and outer circle of the arena in response to a whine and a whine plus chuck.

86   Chapter Five

Increased localizability of calls with chucks also predicts that a female would be attracted preferentially to a call with more chucks compared to a call with fewer chucks. I tested the preferential response of females to a whine plus chuck versus a whine plus 2 chucks in the test arena I used for female-choice tests. The females did not discriminate among these calls (7 responses to whine plus 2 chucks, 6 responses to whine plus 1 chuck; $P = 0.50$).

I also attempted to test the accuracy of the female's ability to localize simple and complex calls from a relatively far distance. In a large dark room on Barro Colorado Island, I painted a 5-by-5-meter grid on the floor. A female was placed at one end of the grid, and at the other end was a speaker that continuously broadcast either a whine or a whine plus 3 chucks (fig. 5.16). I had planned to videotape the females and compute their jump angles as they approached the speaker broadcasting each call type. However, the females usually would not approach the speaker under the high light levels required for the operation of the video camera. The preliminary results of these experiments left me with the impression that a female approached a whine as directly as she approached a whine plus 3 chucks (fig. 5.16).

The above experiments all rejected the hypothesis that females are attracted preferentially to complex calls because these calls are easier to locate. I suggest, however, that this possibility still should not be dismissed. These experiments were conducted under controlled conditions, and it was

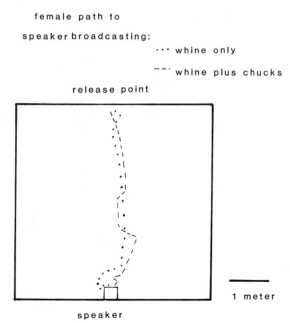

Figure 5.16 The path of a female as she approached a speaker broadcasting a whine and a whine plus 3 chucks.

especially important that they were conducted in the absence of background calling by other males. Perhaps calls with chucks would stand out (acoustically) against the background of a chorus better than calls without chucks. C. Brown (1982) also warns against drawing conclusions from studies of localization abilities that are conducted in the absence of background noises normally present in the environment. I was not able to test this hypothesis, so we are left with no evidence supporting the contention that females can locate a male producing complex calls more readily than a male producing simple calls. Because of the experimental difficulties in conclusively rejecting the null hypothesis, I suggest that the question remains open.

Females prefer calls with chucks over calls without chucks. In the field females appear to select mates, and larger males are more likely to mate. These statements raise an interesting series of questions about female mate choice: Are these phenomena related? Do larger males produce chucks that are measurably different from the chucks of smaller males? If so, do females respond preferentially to these differences in such a way that they are more likely to select larger males as mates?

A general trend exists both among (Martin 1972; Duellman and Pyles 1983) and within (Zweifel 1968; Davies and Halliday 1978; Gerhardt 1982) species of frogs such that larger males produce calls with lower frequencies, although this trend is not true for all species (Zweifel 1968). The reason is no great mystery, for it is predicted by elementary physics. The fundamental frequency of any vibrating membrane will decrease as its mass increases (Roederer 1975). The fundamental frequency of a male's call is determined primarily by the mass of its vocal cords (Martin 1972). Vocal cord mass increases as a function of male size, and consequently, the frequency of the advertisement call decreases as a function of male size. This course of logic suggests that females might preferentially select larger males as mates because they are preferentially attracted to calls with lower frequency chucks.

During the preliminary study in 1978, I recorded the advertisement calls of 32 males of known size (fig. 5.17). When I returned to Cornell University I examine various spectral properties of the call to determine whether any call properties were correlated with male size. I examined the initial and final frequency of the whine in simple calls, the frequency sweep of the whine (initial frequency minus the final frequency), and the fundamental frequency of the chuck. Only the fundamental frequency of the chuck was significantly correlated with male size; the correlations for the other call properties and male size were not statistically significant (table 5.6). In 1979 I recorded the calls of another 100 marked individuals. The data from 1978 and 1979 show that the correlation between male size and the fundamental frequency of the chuck is -0.53 ($P < 0.01$; fig. 5.18). The calls were recorded from males that ranged from 26 to 33.5 millimeters in size, and the chucks had fundamental frequencies that ranged from 200 to 270 Hz. The size range of males recorded, and presumably the frequency range of the chuck, did not encompass the entire range of the male population, which was 24 to 34 millimeters.

Table 5.6 Correlation of Male *Physalamus pustulosus* Size and Various Call Characteristics

Male Size vs.	r	P
Initial frequency of whine	−.11	>.05
Final frequency of whine	−.06	>.05
Frequency sweep of whine	−.10	>.05
Number of chucks	.09	>.05
Fundamental frequency of chuck	−.53	<.01

Note: r correlation coefficient.

The chuck contains some information about male body size, although the predictability of a male's size based on the frequency of his chuck is low (coefficient of determination, $r^2 = .28$). This observation suggests that among certain sizes, females might be selecting larger males due to a preferential response based on differences in the frequencies of the chuck. This response could be based on differences among the fundamental frequencies of chucks, or on one or more of the 14 harmonics of the fundamental frequency, since the harmonics are integer multiples of the fundamental frequency. At this point, the evidence suggesting the mechanism of female mate choice is correlative, but many studies of factors influencing female choice, whether traits of the male or traits of the territory, end here (but see Pleszczynska 1978). However, the ability to investigate preferential phonotaxis of females in choice tests allowed me to test experimentally the hypothesis that females are attracted preferentially to calls with lower frequency chucks.

I could have tested this hypothesis by presenting females with calls from a large male and small male. However, I specifically wanted to examine the role of the chuck in female choice. Although other call properties did not vary predictably with size, they did vary, and I had no idea whether this variation also might influence female choice. I wanted females to choose among calls that were identical except for the frequency of the chuck, and this would not be possible using natural calls. Therefore, I used calls that were produced synthetically and mimicked the natural whine-plus-chuck call (fig. 5.19). In these calls the whines were identical, but the fundamental frequency of the chuck, and consequently all of the harmonics of the chuck, were varied.

To examine female preference based on differences in the frequencies of the chuck, a female was placed in the center of an arena similar to that used by Rand when he examined female preference for simple and complex calls. The arena had burlap walls and contained two speakers, each on an opposite side of the arena and facing towards the arena's center (fig. 5.20). A different call was broadcast from each speaker. The calls alternated in time, and each call had the normal repetition rate for a calling túngara frog. Female preference was determined by noting which speaker the female approached and contacted.

Figure 5.17 The author preparing to record frog vocalizations.

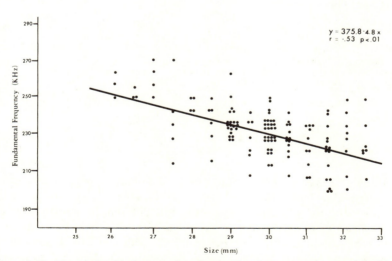

Figure 5.18 The relationship between size of male *Physalaemus pustulosus* and the fundamental frequency of the chuck. From M. J. Ryan, "Female mate choice in a neotropical frog," *Science* 209 (1980): 523–25. Copyright 1980 by the American Association for the Advancement of Science.

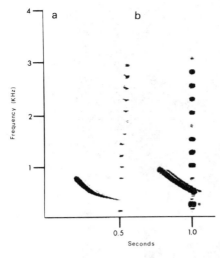

Figure 5.19 Sonograms of a natural túngara frog call (*a*), and a synthetic call (*b*) used to test female preference for differences in the frequencies of the chuck. From M. J. Ryan, "Female mate choice in a neotropical frog," *Science* 209 (1980): 523–25. Copyright 1980 by the American Association for the Advancement of Science.

The first pair of calls I tested had chucks with fundamental frequencies of 200 Hz and 260 Hz, which are within the range of chuck frequencies produced by males in the population. Eight females were given a choice between these two calls, and all of them were attracted preferentially to the 200 Hz call (table 5.7), demonstrating that variation in the calls among male túngara frogs is biologically meaningful. In some comparisons, females are attracted preferentially to chucks with lower frequencies; this result suggests that preferential phonotaxis by females explains some of the size-biased mating success of males.

Table 5.7 Female *Physalaemus pustulosus* Response to Chucks of Different Fundamental Frequencies in Paired-choice Experiments

Test	Frequency	# Females	Frequency	# Females	P
A	200 Hz	8	260 Hz	0	.004
B	210 Hz	9	250 Hz	2	.032
C	220 Hz	10	250 Hz	6	.237

Source: Ryan 1983b.

I was interested in the extent to which differences among chucks resulted in preferential phonotaxis by females, and I was able to test this by presenting females with pairs of chucks that were less different in their fundamental frequencies than 200 Hz and 260 Hz. The females demonstrated that they also behaviorally discriminate between calls with chucks that have fundamental frequencies of 210 Hz and 250 Hz, again preferring the call with the lower frequencies. However, in the simultaneous-choice test using calls with chucks of 220 Hz and 250 Hz the females showed no statistical preference (table 5.7).

Advertisement Call Function   91

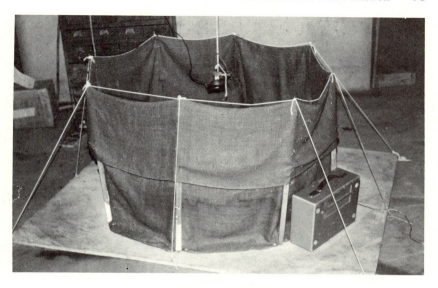

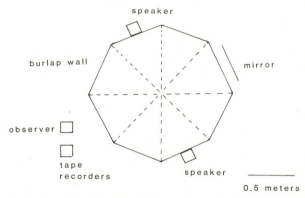

Figure 5.20 A photo and diagram of the arena used for the paired-choice experiments to test female preference among calls.

Statistical testing only allows the rejection of the null hypothesis, not its acceptance (Sokal and Rohlf 1969). Therefore, the above results show that females do respond preferentially to the call with the lower frequency chuck when the calls are at least as different as 210 Hz and 250 Hz. The testing does not demonstrate that females cannot behaviorally discriminate the difference between 220 Hz and 250 Hz, but it does not support the hypothesis that females can make this discrimination. Also, because the females preferentially respond to a call with a chuck that has a 30-Hz-lower fundamental frequency, this does not necessarily mean that females can discriminate this frequency difference among all pairs of calls (e.g., 200 Hz versus 230 Hz).

In a simultaneous-choice test a female is located exactly half-way between two speakers when she is initially presented with the calls being tested. This spatial relationship is quite different from the relationships in the chorus when she samples mates (see chap. 3, especially fig. 3.11). Gerhardt (1982) has presented some important data that call for a careful interpretation of the results of playback experiments. In choice tests using two speakers, female green treefrogs (*Hyla cinerea*) preferred calls that had a low frequency peak of 700 Hz over calls with a 900 Hz low frequency peak. But in experiments in which females were presented with four different low frequency peaks from four speakers, they no longer exhibited this preference. These results are important because they indicate that two-choice playback experiments do not necessarily indicate a female's response in other situations. A two-choice experiment is artificial, as Gerhardt points out, but a four-speaker test might not represent a substantially more realistic approximation of the natural situation in which a female samples males. This certainly is true for female túngara frogs (fig. 3.11); the concordance between the experimental situation and what actually occurs in the field depends on the species being studied.

Due to the frequency filtering of the frog's peripheral auditory system, some frequencies will be perceived by the frog as louder than other frequencies even if the frequencies are produced at the same intensities (see fig. 1.1). For a frog, the perceived loudness of any frequency is dependent on the filtering properties of the auditory system and the absolute intensity of the sound when it reaches the frog. The intensity of the sound at the receiver is dependent on the intensity at which the sound is produced at the source and the distance from the source of the sound to the frog receiving the sound. Some researchers have suggested that female preference for a certain frequency is merely a product of the frequency tuning of the peripheral auditory system. If this were true, then when a female enters a chorus and is equidistant between two males producing calls of the same intensity but different frequencies, she will perceive as louder the calls with frequencies that best match the frequency tuning of her peripheral auditory system. When the female enters a chorus, to use the most simple example, she might be very close to a male producing a "suboptimal" frequency (relative to the tuning of her auditory system) and farther from a male producing an "optimal" frequency. In this case the female might perceive the suboptimal call as louder. Therefore, playback experiments in which a female is equidistant from speakers playing calls of identical intensity are unrealistic. However, assuming that a female automatically responds to the first call that meets some "optimal" criterion of perceived loudness alone is also unrealistic, as is the assumption that she is unable to sample males in a more sophisticated manner.

The observations I reported in the previous chapter clearly suggest that the movements of túngara frog females prior to mating afford them ample opportunity to sample mates. Similar observations have been reported for a number of other species of frogs (e.g., bullfrogs by Emlen 1976; toads by

Sullivan 1983). Prior to mating, a female túngara frog might approach to within 10 centimeters or less of a calling male before moving on to another male. At this distance the call is so intense that, depending on the thresholds of the auditory fibers (e.g., fig. 8 in Capranica and Moffat 1983), the female's auditory system might be saturated. In other words, frequencies that are at the peak of the tuning curve, as well as frequencies on either side of the peak, might be perceived at their maximum loudness. Our observations of túngara frog females suggest that females do not choose males simply and immediately based on some preferential perceived loudness: if this were so a female would not move on to sample other males. These observations of female choice are consistent with theoretical proposals (Janetos 1980) and empirical findings (Downhower and Brown 1980) that suggest females compare males prior to mating.

Other studies have also indicated that extrapolations from data on the neurophysiology of the auditory system to behavioral sensitivity should be made with caution. Zelick and Narins (1982) demonstrated that Puerto Rican treefrogs, *Eleutherodactylus coqui*, do not respond behaviorally to frequencies to which, neurophysiologically, they are very sensitive. Alternatively, Ramer, Jennson, and Hurst (1982) showed that frequencies that should inhibit a neurophysiological response in green frogs, *Rana clamitans*, actually elicit an increased behavioral response. Capranica and Moffat (1983) have emphasized the problems of extrapolating neurophysiological findings across species. But is should be remembered that although the playback experiments and measures of neurophysiological sensitivity do not always predict the animal's behavior in nature, such observations provide information on the sensory capabilities of the animal, and these findings must simply be interpreted with care.

The results presented thus far show that larger males produce calls that have chucks with lower fundamental frequencies, and that in some cases females are attracted preferentially to calls with lower frequency chucks. This preference, therefore, explains some of the size-biased variation in male mating success observed in the field. However, one question that demands attention is how reliably the frequency of the chuck predicts male size. The answer is simple—not reliably at all. It explains only 28% of the variation in male size.

This question can be framed in a manner more amenable to analysis: If a female chose the call with the lower frequency chuck in a pair of calls, how often would she actually select the smaller male; that is, how often would she "make a mistake"? (This term is used for convenience only and does not imply that the preferential phonotaxis of females indicates that they are attempting to select larger males.) Females discriminate between calls with fundamental frequencies of 200 Hz and 260 Hz. Referring to figure 5.18, we see that if a female compared all of the calls with chucks that had fundamental frequencies lower than or equal to 210 Hz with calls that had chucks of fundamental frequencies greater than or equal to 260 Hz, 60 possible pairwise comparisons would result. If in each comparison a female chose the call with the lower frequency chuck, she also would always select the larger male. Her probability of "making a mistake" would be 0, because never is a male with a lower frequency chuck (in this case, less than or equal to 200 Hz) smaller than a male with a higher frequency chuck (greater than or equal to 260 Hz). Females also chose the lower frequency chuck when given a choice between calls that had chucks with fundamental frequencies of 210 Hz and 250 Hz. In this comparison, figure 5.18 shows that there are 198 possible pairwise comparisons between males with calls that have chucks with fundamental frequencies less than or equal to 210 Hz, and males with chucks having fundamental frequencies greater than or equal to 250 Hz. If in these comparisons a female always chose the lower frequency chuck, she actually would choose the smaller male with a probability of 0.06. Examining the size of males with chucks having fundamental frequencies of less than or equal to 220 Hz and those greater than or equal to 250 Hz—the level at which females did not show a preference—a female always choosing the lower chuck would "make a mistake" with a probability of 0.09 in the 264 possible pairwise comparisons.

Halliday (1983a), citing a yet-to-be-published paper by Doherty and Gerhardt, stated that the results of my female-choice tests with *P. pustulosus* could be criticized because the choice presented to females is between extremes. He stated that to demonstrate female choice one must show that females prefer calls of larger males to those of average-sized males. Problems with this criticism are as follow: My contention is that a differential response by females based on frequency differences in calls among males tends to enhance the mating success of larger males, not that females are able to accurately measure male size using call characters. Evidence based on patterns of female movements prior to amplexus, size-biased male mating success, the correlation between male size and call frequency, preferential female phonotaxis, and as I will discuss later in this chapter, phylogenetic analyses of call evolution all taken together support the hypothesis that female choice influences male mating success in this species. Halliday's criticism implies that female choice is not an important selective force because it is not perfect; that is, in some comparisons in the playback experiments choice is not exhibited and thus selection does not act. Although the playback experiments suggest the sensory limits of discrimination, they do not indicate how often in the field females can select mates, given what can be their very extensive movements (sampling?) among males prior to initiation of mating. Furthermore, Halliday's comments imply a very rigid definition of selection. He seems to view it as a perfecting force rather than a biasing force, a view that is at variance with most other concepts of selection (e.g., Wade and Arnold 1980). For example, consider the following hypothetical situation. Within a population of a species of predator, there is a range in maximum running speed from 20 to 30 miles per hour. The fastest animals are more successful at capturing prey than are the slowest, and the latter usually leave few or no offspring because they starve. However, there is no detectable difference in the ability of individuals with maximum speeds of 25 and 30 miles per hour to capture prey. Still, I would argue that evidence exists for selection on running speed for this species of predator. Similarly, even though females do not always select the males with the lowest frequency calls, the fact that this does occur over a certain range of frequencies suggests selection on call frequency.

Without necessarily assuming a cause and effect relationship, I conclude that females cease to exhibit a preference between chuck frequencies as this comparison becomes less likely to result in choice of larger males. Therefore, in comparisons between certain frequencies, females preferentially respond to calls based on frequencies of the chucks in such a way that larger males are more likely to be selected as mates. It is important to resist the tendency to invert this statement. That is, the observation that preferential phonotaxis by females results in larger males being selected as mates *does not* allow us to conclude that that females exhibit preferential phonotaxis *in order* to select larger males as mates. (Although the telelogical approach has heuristic value, in studies of sexual selection this approach is often translated into conclusions regarding cause and effect.)

The Chuck and Sexual Selection by Male Competition

The previous experiments show that the frequency structure of the chuck influences female behavior. I also conducted experiments to determine whether chuck frequency influenced male behavior. First, I presented to a male calling in the pond behind Kodak House synthetic calls that had chucks with fundamental frequencies of either 200 Hz or 260 Hz. The stimuli were presented for 30 seconds each with an intervening 1 minute no-stimulus period. This procedure was conducted 4 times. An analysis of variance shows that there were no significant differences in the number of whines ($F = 0.06$, $P > 0.10$) or in the number of chucks ($F = 0.01$, $P > 0.10$) produced by the male in response to the two stimuli (table 5.8).

Table 5.8 Male *Physalaemus pustulosus* Response to Chucks of Different Fundamental Frequencies

	200 Hz		260 Hz	
Test	Whines	Chucks	Whines	Chucks
1	9	16	14	22
2	5	6	2	3
3	4	7	5	7
4	3	4	3	3

A second test is provided by data that were collected in experiments originally designed to investigate the separate roles of the whine and the chuck in eliciting male vocal responses. These experiments will be discussed in detail later in this chapter, but the data relevant to the hypothesis that chuck frequency influences male behavior will be considered here.

In these experiments males of various sizes were presented with a natural whine-plus-chuck call, and the males' response was recorded. The chuck had a fundamental frequency of about 230 Hz. If the frequency of the chuck, which is related to male size, influences male behavior, one might expect that a male's response to a given call will be influenced by his size relative to the size of the male producing the call. This expectation has been found to be true in green frogs (Ramer, Jennson, and Hurst 1983). The results of experiments with túngara frogs show that there is no correlation between a male's size and the number of whines ($r = -0.22$, $P > 0.10$, $N = 18$) or the number of chucks ($r = 0.16$, $P > 0.10$) produced in response to the stimulus call. Also, visual inspection of the data do not suggest any pattern of call response that is biased by male size. Although in this species the frequency of the chuck influences female behavior, it does not appear to influence male behavior.

Sexual Selection and Call Evolution: A Phylogenetic Approach

Since females prefer chucks with lower frequencies, we must ask if this selection exerted through female choice actually has had an evolutionary effect on the male call. Have male túngara frogs evolved lower frequency calls? A logical scenario might suggest that if a mutation occurred that resulted in males having calls with low frequencies for their size, then under the influence of female choice this mutation should spread through the population. Females would still prefer lower frequency calls, but larger males would not be chosen as mates. However, as this mutant allele (or complex of alleles) increased in frequency, male size and call frequency eventually would again become correlated. The "mutant" males, which would now comprise a majority of the population, would still have lower calls, but regardless of the frequency of a male's call, it should become lower as he, and his vocal cords, grow. Female preference for lower frequency calls would again result in the choice of larger males as mates. Therefore, as pointed out by West Eberhard (1979), oscillations might occur in the correlation between a male trait (e.g., size) and the display (e.g., call frequency) used to choose mates.

The above scenario is consistent with selection theory, but has it occurred? It is hard to say. As will be discussed in detail in the next chapter, if females of any species gain a nongenetic benefit from a mate (e.g., better territory, uninterrupted matings, more eggs fertilized), and the males in some way indicate the magnitude of this benefit in their displays, then evolution of female choice will result from a nongenetic advantage obtained by discriminating females. This will be true whether or not the male trait being selected has heritability. If a female chooses a male with a better territory, she will benefit whether or not her sons are able to differentially acquire better territories. If the male trait being selected has no heritability, then that male trait will not evolve in response to sexual selection. Therefore, selection can influence the evolution of the female preference without influencing the evolution of the male trait on which females base their choice. If the heritability of the chuck's frequency is zero, we would not expect the frequency of the túngara frog call to have decreased over evolutionary time.

Do sexual displays have high heritability? A strong genetic basis to the anuran advertisement call probably exists. For example, Burger (1980, cited in Gerhardt 1982) raised two species of hylid frogs, *Hyla cinerea* and *H. chrysoscelis*, in the acoustic environment of either heterospecific or conspecific calls. Males always produced the conspecific call. However, heritability is dependent not only on whether call characteristics are inherited (i.e., genetically transmitted from one generation to the next), but also on how much of the variation of call characteristics among males results from additive genetic variance (chap. 1; Falconer 1981). Evidence exists for significant heritabilities for the sexual display behavior of some organisms. For example, Cade (1981) showed that the amount of calling by male field crickets influences their ability to attract mates, and selection for low and high levels of calling have heritabilities of 0.50 and 0.53, respectively. MacDonald and Crossley (1982) also have shown that reproductive behaviors influencing male mating success in *Drosophilia melanogaster* respond to selection. Unfortunately, no data allow us to estimate the heritability of the frequency of the call of any frog, or for that matter, of any other anuran trait related to reproductive behavior.

Testing the hypothesis that the frequency of túngara frog calls has decreased throughout evolutionary time is a difficult task. Even if an adequate fossil record existed for this species, behaviors are notorious for their failure to fossilize. However, some data exist that bear on this hypothesis; these data concern call relationships among closely related species. I offer this information not because it provides a definitive test of the hypothesis, but because the comparative, phylogenetic approach is the only option we have for exploring the evolutionary history of animal behavior (Lorenz 1981; Mayr 1982). The comparative approach, especially when it encompasses some of the more explicit techniques for comparisons, can be a powerful tool, and one that I think has not been appreciated by many behaviorists.

Among closely related frogs in which call frequency has been determined, a correlation usually exists between male size and the dominant frequency of the advertisement call (e.g., the genus *Bufo*—Martin 1972; the family Hylidae—Duellman and Pyles 1983). Everything else being equal, if female choice has resulted in the evolution of low frequency túngara frog calls, then the frequency of the call should be lower than that predicted by the size-frequency relationship of closely related frogs. I offer this hypothesis cautiously because it is not likely that everything else is equal. For example, if females preferred lower frequency calls in related species, then frequency of the túngara frog call might be accurately predicted by the general size-frequency relationship. This would lead to the rejection of the hypothesis that female choice has influenced call frequency, even if it had had such an effect. Not surprisingly, nothing is known about the mating systems of close relatives of *P. pustulosus*. Another problem is that if the frequency of the túngara frog call was lower than predicted, this could be due to causes other than

Table 5.9 Lowest Frequency Present in the Advertisement Call and Typical Male Size, Thirty Species of Leptodactylinae

Species	Lowest Frequency (Hz)	Male Size (mm)	$d^2 L\ (+/-)$	$d^2 P\ (+/-)$
Physalaemus pustulosus	230[e]	29.5[e]	0.63 (−)	0.44 (−)
Physalaemus albonotatus	800[c]	30.0[d]	0.001 (−)	0.01 (+)
Physalaemus biligonigerus	400[c]	35.5[c]	0.11 (−)	0.01 (−)
Physalaemus cuvieri	600[c]	30.0[c]	0.04 (−)	0.004 (−)
Physalaemus fernandezae	3100[c]	20.0[c]	0.11 (+)	0.22 (+)
Physalaemus gracilis	470[b,e]	27.0[f]	0.13 (−)	0.04 (−)
Physalaemus henseli	457[c]	20.0[d]	0.25 (−)	0.13 (−)
Physalaemus maculiventris	1100[b,e]	17.0[f]	0.03 (−)	0.001 (+)
Physalaemus olfersi	1300[b,e]	34.0[f]	0.03 (+)	0.15 (+)
Physalaemus riograndensis	1000[c]	25.0[d]	0.004 (−)	0.01 (+)
Physalaemus santafecinus	400[c]	34.0[c]	0.12 (−)	0.14 (−)
Physalaemus signiferus	700[c]	27.0[c]	0.18 (−)	0.001 (−)
Adenomera hylaedactyl	3700[a]	22.7[b]	0.21 (+)	
Adenomera marmorata	2200[a]	20.8[b]	0.04 (+)	
Leptodactylus bolivianus	200[a]	94.0[b]	0.31 (+)	
Leptodactylus bufonius	100[a]	51.6[b]	0.06 (+)	
Leptodactylus chaquensis	250[a]	94.0[b]	0.02 (−)	

Table 5.9, *continued*

Species	Lowest Frequency (Hz)	Male Size (mm)	$d^2 L (+/-)$	$d^2 P (+/-)$
Leptodactylus fuscus	800[a]	42.8[b]	0.002 (+)	
Leptodactylus gracilis	500[a]	43.0[b]	0.02 (−)	
Leptodactylus labialis	600[a]	34.7[b]	0.0004 (+)	
Leptodactylus latinasus	3000[a]	31.2[b]	0.25 (+)	
Leptodactylus melanonotus	2000[a]	46.0[b]	0.23 (+)	
Leptodactylus mystaceus	2200[a]	53.0[b]	0.33 (+)	
Leptodactylus ocellatus	100[a]	140.0[b]	0.13 (−)	
Leptodactylus pentadactylus	200[a]	137.2[b]	0.07 (−)	
Leptodactylus podicipinus	500[a]	38.0[b]	0.04 (−)	
Leptodactylus poecilochilus	350[a]	44.8[b]	0.08 (−)	
Leptodactylus syphax	1500[a]	70.0[b]	0.28 (+)	
Leptodactylus wagneri	1000[a]	63.0[b]	0.10 (+)	
Vanzolinius discodactylus	2580[a]	35.0[b]	0.23 (+)	

Sources: a, Straughn and Heyer 1976; b, Heyer 1980; c, Barrio 1965; d, Cei 1980; e, author's analysis; f, Lynch 1970.

Notes: d^2_L is the square of the departure of the logarithm of the observed frequency from the frequency expected from the regression of frequency on size for the subfamily (regression equations are given in table 5.10). (+) indicates that the observed frequency was higher than predicted and (−) indicates that the observed frequency was lower than predicted. d^2_P is determined as is d^2_L, but the analysis is restricted to members of the genus *Physalaemus*.

female choice, such as the influence of the environment on call structure (e.g., Morton 1975).

Given the above caveats, I examined the relationship between call frequency and size for species closely related to *P. pustulosus*. The family to which *Physalaemus* belongs, the Leptodactylidae, has more species than most other vertebrate families, and the phylogenetic relationships among members of this family are still not clear (Lynch 1971; Heyer 1975). Therefore, for this analysis I considered members of the subfamily Leptodactylinae (fig. 2.5). For the 30 species analyzed, the lowest frequencies of their calls, and their sizes are listed in table 5.9. The fact that the size of the individual producing the call analyzed is not known undoubtedly introduces added variance into the size-frequency relationship. However, this variance would tend to obscure rather than to enhance any trend in the general size-frequency relationship.

Some of the species listed (table 5.9) produce narrowly tuned calls, while other species produce calls that cover a broad range of frequencies. Because this analysis addresses the question of whether *P. pustulosus* produces an exceptionally low frequency call for its size, I use the lowest frequency in the call of each species in the analysis. The logarithm of the call's lowest frequency (log Hz) was regressed on the logarithm of species size (log SVL) using the least-squares method (fig. 5.21). This analysis shows that the difference in size among species explains a significant amount of the variation of the lower frequencies of their calls (table 5.10).

Table 5.10 Logarithm of Call Frequency Regressed on the Logarithm of Body Size for Various Taxonomic Groups

Taxon	Regression Equation	F	N	P
Subfamily Leptodactylinae	$Y = 4.38 - 0.94 X$	14.4	29	$< .005$
Subfamily Telmatobiinae	$Y = 4.33 - 0.61 X$	11.3	15	$< .01$
Genus *Physalaemus*	$Y = 4.65 - 1.25 X$	3.5	11	.09
Genus *Leptodactylus*	$Y = 4.26 - 0.77 X$	4.3	15	.06

Note: The data for *P. pustulosus* were not used in determining the regression equation for the subfamily Leptodactylinae or the genus *Physalaemus*.

The data graphically presented in figure 5.18 show that size constrains call frequency to a limited degree, but much less so in the genus *Physalaemus* than it does in the genus *Bufo* (Martin 1972) or the family Hylidae (Duellman and Pyles 1983). Many species have calls much lower or higher than that predicted by the regression of frequency on size, and among the 12 species of *Physalaemus* analyzed (fig. 5.21), size does not explain a significant portion of the variation in frequency (table 5.10). Although the evolutionary importance of male size in determining the spectral properties of the anuran advertisement call has been emphasized (e.g., Blair 1955), the data presented here clearly demonstrate that to some extent, call frequency can evolve independently of male body size.

The ranges of male size and the lower frequency of túngara frog calls are shown in figure 5.12. Visual inspection of these data suggest that the lower frequency of the call of *P. pustulosus* is much lower than that predicted by the size-frequency relationship of closely related frog species. For example, *P. pustulosus* and *Leptodactylus pentadactylus* have almost identical low frequencies (table 5.9). As is seen in figure 5.22, the animals are of distinctly disparate body size.

I quantified the degree to which the call of a species departs from the expected frequency predicted by the regression for members of the subfamily Leptodactylinae by calculating the square of the difference between the logarithms of the observed frequency and the expected frequency from figure

Advertisement Call Function 101

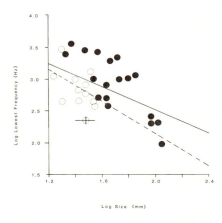

Figure 5.21 The relationship between the logarithm of male size (log SVL) and the logarithm of the lowest frequency in the advertisement call (log Hz) for 29 species of the subfamily Leptodactylinae. Open circles represent members of the genus *Physalaemus* and closed circles represent other members of the subfamily (data from table 5.8). The solid line represents the regression of frequency on size obtained for all members of the subfamily, and the dashed line represents the regression for only members of *Physalaemus*. The circle intersected by the vertical and horizontal lines represents the midpoint and the ranges of the size and frequency for *P. pustulosus* (data from fig. 5.15).

5.21. These data (d^2) indicating the departures from the regression, as well as the direction of the departures, are presented in table 5.9 and figure 5.23. Using the midpoint of the size and frequency range for *P. pustulosus* (fig. 5.18), we see that the observed frequency departs from the predicted frequency by 252% more than do the frequencies of the calls of any other species of the subfamily (fig. 5.20).

The same analysis was conducted with only species of *Physalaemus*. However, since the regression of frequency on size does not explain a significant amount of the variation in call frequency (table 5.10), these results must be considered with caution. In the within-genus comparison, *P. pustulosus* still has a call with a frequency much lower for its size than that of any of its congenerics; the *P. pustulosus* call departs from this regression 314% more than that of the calls of any other species of *Physalaemus* (fig. 5.23).

Figure 5.22 *Physalaemus pustulosus* and the much larger *Leptodactylus pentadactylus*.

102   Chapter Five

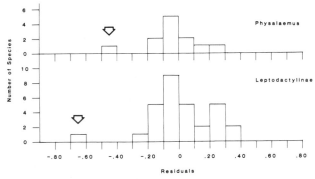

Figure 5.23 Frequency distribution of the magnitude and the direction of the departure of the observed low frequency of the call from the expected based on the size-frequency relationships of the subfamily Leptodactylinae and the genus *Physalaemus*. The arrow indicates the departure exhibited by *P. pustulosus*.

The comparison of size-frequency allometries among closely related species does show that the frequency of the *P. pustulosus* call is low for the size of the frog, but it does not indicate whether this low frequency call is a primitive or a derived character. Although the phylogenetic comparison shows that call frequency can evolve independently of body size, we still must ask if *P. pustulosus* evolved a low frequency call or if the other species analyzed all evolved high frequency calls. Although it seems more likely that the former is true, this question can be addressed in a more rigorous manner using cladistic analysis.

The direction of character evolution, or polarity, is a crucial issue in the reconstruction of phylogenies. In order to determine the sequence in which species diverge it is necessary to know if a character is primitive or derived. In cladistic analysis, one method of resolving polarity is that of outgroup comparison:

> Given two characters that are homologues and found within a single monophyletic group, the character that is also found in the sister [closely related] group is the plesiomorphic [shared primitive] character whereas the character found only in the monophyletic group is the apomorphic [derived] character. (Wiley p. 139, 1981)

Cladists typically use outgroup comparisons to resolve relationships among species. However, as I apply the method here, outgroup comparisons also can provide important information about evolutionary patterns of character evolution.

In applying the definition of outgroup comparisons to the question of call evolution, I suggest that if the size-frequency relationship exhibited by *P. pustulosus* is shared with a closely related group (in this case a closely related subfamily or genus) then the low frequency call of this species probably is not a derived character. Instead, it should be considered plesiomorphic—a primitive character shared by species with a common ancestor—and the alternative hypothesis, that the other members of the group containing *P. pustulosus* have evolved higher frequency calls would be supported.

Based on the available data, I chose the subfamily Telmatobiinae as the closely related subfamily for the outgroup comparison. Data on call frequency were obtained from Drewry and Rand (1983) who analyzed the calls of 14 species of *Eleutherodactylus*. The sizes of these species were provided by Rivero (1978). The genus *Leptodactylus* was used as the closely related genus, and data on call frequency and body size are in table 5.9. The results of the regression of frequency on size for the outgroups are presented in table 5.10, and the resulting regression equations are in figure 5.24.

P. pustulosus has a call frequency that is low for its body size relative to other members of the subfamily Leptodactylinae (figs. 5.21, 5.23), and the size-frequency relationship exhibited by *P. pustulosus* clearly is not shared by members of the subfamily Telmatobiinae (fig. 5.24). The analogous results are obtained for the outgroup comparison at the genus level. The *P. pustulosus* call has a low frequency relative to other members of its genus (figs. 5.21, 5.23), and the call frequency of this species is also low when compared to members of the closely related group *Leptodactylus* (fig. 5.24). This latter analysis should be considered with caution. As mentioned previously, size does not explain a significant amount of the variation in call frequency within the genus *Physalaemus*, which is also true for the genus *Leptodactylus* (table 5.10). Nevertheless, outgroup comparisons at both the levels of the subfamily and the genus support the intuitive notion that the relatively low frequency call of *P. pustulosus* is a derived character.

If female choice has influenced the evolution of the spectral properties of the call of *P. pustulosus*, males should have evolved lower frequency calls,

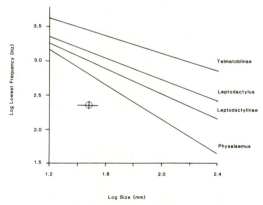

Figure 5.24 Regression lines of the logarithm of call frequency on the logarithm of body size for species in the subfamilies Leptodactylinae and Telmatobiinae, and the genera *Physalaemus* and *Leptodactylus*. Regression equations are represented in table 5.10. The circle represents the mean body size and call frequency for *P. pustulosus* and the lines through the circle represent the ranges.

and that is what these results suggest. When male size and spectral characteristics of the túngara frog call are compared to those of closely related species, the túngara frog is observed to produce calls that contain significantly lower frequencies than would be expected for its size. Outgroup comparisons suggest that the lower frequency call is the derived state; thus, *P. pustulosus* evolved a low frequency call, as opposed to the other species analyzed evolving high frequency calls.

In *P. pustulosus*, the exceptionally low frequency of the call is due to the production of the chuck. If males produced only whines, their calls still would be low for their size, but not as strikingly so. Only the chuck appears to be influenced by sexual selection. As I will discuss later in this chapter, the whine functions in species recognition, and the chuck makes no contribution to the ability of the female to discriminate among species. Gerhardt (1982; see also Fisher 1958) has suggested that in some frogs, species discrimination will constrain the manner in which mating calls can evolve; a call will not evolve in such a way that it will reduce the ability of the female to choose the correct species as a mate. In *P. pustulosus* the chuck is able to evolve without any such constraints since it plays no role in species recognition. Also none of the other species analyzed has a call with two distinct components—only túngara frogs have a call component that functions only in intraspecific mate choice. Thus the chuck might exhibit more of an evolutionary response to sexual selection than call characters of other frogs. These facts are all consistent with the interpretation that sexual selection has influenced the evolution of male vocalizations, which suggests that female choice has resulted in the evolution of the chuck per se, as well as in the relatively low frequencies of the chuck. Habitat effects are probably not responsible for the observed differences among frequencies, since *P. pustulosus* breeds in the same habitats as many of the species listed in table 5.8.

A Morphological Correlate of Call Evolution

The previous discussion suggests that male túngara frogs have relatively low frequency calls for their size. In part, this is due to the fibrous masses that produce the chuck. The spectral properties of anuran vocalizations, as mentioned previously, are determined primarily by the vibration of the vocal cords (Martin 1972), and the fundamental frequency of the vibration decreases as the mass of the vocal cords increases (e.g., Roederer 1975). But other factors, such as the size and shape of the vocal tract and the vocal sac, also influence the spectral structure of the call. A good example of how these structures might alter the call spectrum comes from human speech (reviewed by Rossing 1982). When human vocal cords (or any other membrane) vibrate, the relative amplitude is greatest in the fundamental frequency, and amplitude decreases with higher harmonics (fig. 5.25*A*). But the relative amplitude-frequency spectrum of human speech is quite different from the

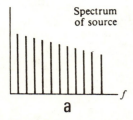 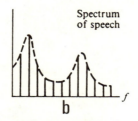

Figure 5.25 *A*, the power spectrum of the vibration of the human vocal cords, and *B*, of typical human speech. The difference between the spectra is due to the resonating and radiating properties of the human vocal tract (from Rossing 1982).

spectrum of the vocal cord vibration (fig. 5.25*B*) because the human vocal tract acts as a resonator and a radiator, thus altering the spectral structure of the vocal cord vibrations before they are emitted as speech. The degree to which the spectrum at the source of vibration (i.e., the vocal cords) is altered depends on the shape and size of the vocal tract.

The frog's vocal sac probably functions as a sound radiator (Martin 1972), although few species have been examined (Capranica and Moffat 1980). Above a certain frequency known as the cutoff frequency, any radiator (biological or mechanical) will radiate all frequencies at approximately the same efficiency. Below the cutoff frequency, the efficiency of the radiator decreases drastically (Beranek 1954). The cutoff frequency is a function of the size of the radiator; larger radiators will have lower cutoff frequencies. This is because frequency and wavelength are inversely related (wavelength = speed of sound/frequency), and a larger radiator transmits longer wavelengths more efficiently than does a smaller radiator. Specifically, a radiator will transmit sounds at maximum efficiency when $C/i > 1$, where C is the circumference of the radiator and i is the wavelength (Beranek 1954). Because $C = 2\pi r$, where r is the radius of the radiator, and $i = v/f$, where v is the velocity of sound (about 330 meters/second under "normal" conditions), and f is frequency, then the cutoff frequency (f_c) can be determined empirically: $f_c = v/(2\pi r)$. To maximally transmit the fundamental frequency of the chuck, f_c would have to be less than 200 Hz and $r > v/2\pi f_c$, or 26 centimeters. The average size of a male túngara frog is about 3.0 centimeters. The cutoff frequency for a male of this size is 3500 Hz, which comes close to predicting the peak energy concentration of the chuck (fig. 5.6*B*). This demonstrates that vocal sac size can be a limiting factor in the transmission of the low frequency components of frog calls.

For its size, *P. pustulosus* has an enormous vocal sac (fig. 5.26) which should aid in, but not maximize, the transmission of the lower frequencies of the chuck. Data presented by Drewry, Heyer, and Rand (1982) also support this interpretation. *P. gracilis* produces an advertisement call similar to the *P. pustulosus* whine. The fundamental frequency of the *P. gracilis* call sweeps from 750 Hz to 450 Hz. The fundamental frequency is the dominant

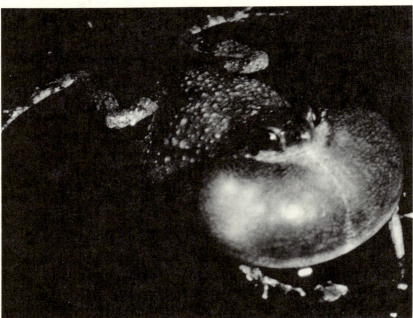

Figure 5.26 *A*, male *Physalaemus pustulosus* prior to call production with the vocal sac deflated, and *B*, during calling with the vocal sac inflated.

frequency at the beginning of the call, but the dominant later shifts to upper harmonics at the end of the call. Nevertheless, substantial energy is contained in the fundamental frequency throughout the call. Like *P. pustulosus*, *P. gracilis* also has a large vocal sac.

Another species, *P. olfersi*, produces a call that, determined from the analysis of the harmonic structure, has a fundamental frequency between 160 Hz and 200 Hz, which is lower than that of either *P. pustulosus* or *P. gracilis*. However, the lowest frequency in the *P. olfersi* call visible on a sonogram is 1500 Hz, approximately the eighth harmonic of the fundamental frequency. Unlike *P. pustulosus* and *P. gracilis*, *P. olfersi* has a relatively small vocal sac. *P. olfersi* appears to be too small to radiate the lower frequency harmonics produced by its vocal cords—an assumption consistent with the interpretation that the *P. pustulosus* vocal sac plays an important role in the transmission of the relatively low frequencies of the chuck. An evolutionary change in the lower frequency limit of a frog call necessitates not only an increase in the mass of the vocal cords, but also an increase in the size of the vocal sac. Because the maximum size of the vocal sac is achieved only when it is fully inflated, lung capacity and the size and power of the multiple groups of muscles associated with breathing and calling (Martin and Gans 1972) might also be affected. The evolution of calls with lower frequencies thus appears to entail a complex series of modifications of the frog's morphology.

Another interesting difference among the above three species of *Physalaemus* is that the larger vocal sacs of some of the species of *Physalaemus* allow males to float on the water's surface while calling ("los que inflados de aire en general provocan la flotacion del ejemplar que canta" [Barrio 1965]). Drewry, Heyer, and Rand (1982) note that both *P. pustulosus* and *P. gracilis* call from small pools of water while floating on the surface, but that *P. olfersi* calls from clumps of grass next to small pools. Thus the differences in vocal sac size among these species influence not only the spectral characteristics of their calls, but might also influence the type of habitat from which a male can call. The evolutionary interpretation, again, is speculative. Is sexual selection ultimately responsible for the large vocal sacs of *P. pustulosus*? If so, did selection for lower frequency calls lead to the evolution of enlarged vocal sacs, which then resulted in males being able to float on the water while calling? Or as unlikely as it seems, was the selection initially made for the ability to float while calling, for which the vocal sacs happened to serve well, and this consequently led to the transmission of lower frequencies that were already being produced by the vocal cords?

Function of the Whine and Chuck

The previous experiments and discussions have focused primarily on the role of the chuck in female mate choice. The fact that females respond to only a whine, but prefer whines with chucks, suggests that the whine and the chuck

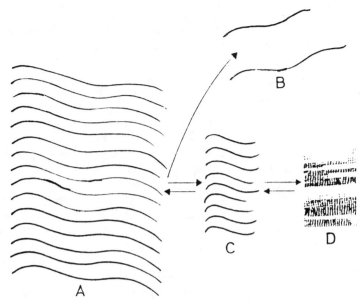

Figure 5.27 Classification of call types of the genus *Physalaemus* (adapted from Barrio 1965).

interact to convey different types of information from the male to the female. I conducted a series of experiments to determine the functional significance and interaction of these two call components.

The advertisement calls of species of the genus *Physalaemus* show considerable variation. In his review of the Argentinian species of *Physalaemus*, Barrio (1965) grouped the known advertisement calls of the genus into 6 types (fig. 5.27). Although the classification did not have a phylogenetic basis, he suggested that calls sweeping upward in frequency (type B) were derived from type A calls. Barrio classified the *P. pustulosus* call as type C, together with *P. cuvieri*, *P. santafecinus*, *P. signiferus*, and *P. aguirrei*. We now know that the túngara frog call is distinguished from the calls of others with this call type, in fact from all other members of the genus, by having a second call component—the chuck. Therefore, I suggest that the *P. pustulosus* call is different than, and might be derived from, a type C call, and I have termed the *P. pustulosus* call as type E (fig. 5.27). Although the figure of the call types (fig. 5.27) seems to suggest an evolutionary pathway by which the calls evolved, little relationship appears to exist between call types and species groups as classified by Lynch (1970; table 5.11). If the species groups truly reflect phylogenetic relatedness, then within certain constraints, the call characteristics found in the genus *Physalaemus* did not evolve in a conservative manner; although this conclusion cannot be determined for certain until the species relationships within the genus *Physalaemus* are

Table 5.11 Call Type and Species Group of Various Species of *Physalaemus*

Species	Call Type	Species Group
biligonigerus (=fuscomaculatus)	A	biligonigerus
albonotatus	A	cuvieri
riograndensis	A	cuvieri
gracilis	A	cuvieri
fernandezae	B	cuvieri
cuvieri	C	cuvieri
santafecinus	C	biligonigerus
signiferus	C	signiferus
aguirrei	C	cuvieri
pustulosus	E	pustulosus
henseli	D	cuvieri
centralis	D	cuvieri
maculiventris	D	signiferus
obtectus	D	signiferus

Sources: Call type, Barrio 1965; species groups, Lynch 1970.

better understood (cf. Cannatella and Duellman 1984). This apparent lack of conservation of form has been predicted for characters that evolve under the influence of sexual selection (e.g., West Eberhardt 1979, 1983; Lande 1982).

A whinelike call appears to be typical of the advertisement call of many members of the genus *Physalaemus*, and the complex advertisement call of túngara frogs appears to be derived from a whinelike call (i.e., type C). Therefore, the whine likely contains information that allows identification of the species by conspecifics. This suggestion is also supported by the observation that although females prefer complex calls, they will respond to a whine in the absence of calls with chucks. I examined the role of the whine in species recognition, and I specifically considered whether the frequency sweep played any role in coding this information. This study was done by comparing the ability of different sound stimuli to elicit vocal responses from males and preferential phonotaxis from females.

Each male was presented with one sound stimulus from a stimulus pair at a rate of 1 stimulus every 2.3 seconds for 30 seconds. During stimulus presentation the male's response was recorded. Then he was presented with the second stimulus of the pair and his response was recorded in a similar manner. I determined which stimulus more effectively elicited a vocal response by comparing the total number of calls, the total number of chucks, the time to the first call (latency to first response), and the time to the first chuck in response to each stimulus of the pair.

In these experiments a whine plus chuck more effectively elicited a vocal response (i.e., more responses and a shorter latency to first response) from males than a chuck only (table 5.12). (A chuck does not occur in nature alone, but was dissected from a natural whine plus chuck.) No difference was found in a male's response to a whine plus chuck relative to a chuck plus whine. I also played the males a reversed whine to determine whether the direction of the frequency sweep was an important feature of the call for eliciting a vocal response. With these calls a male received the same frequencies at the same relative amplitudes as a normal whine, but in the reversed direction. Both a whine and a whine plus chuck more effectively elicited responses from males when they were broadcast in the normal, as opposed to the reverse, direction (table 5.12).

Female choice tests were used to determine the response of females to the same stimuli for which the males were tested. Rand already had demonstrated that a female would not approach a speaker broadcasting only a chuck. As with the males, I showed that the females did not discriminate between a whine plus chuck and a chuck plus whine (table 5.13). And again, the reversed whine was less likely to elicit phonotaxis from females than was a normal whine (table 5.13).

Table 5.12 Evoked Vocal Responses of Male *Physalaemus pustulosus*

			Probabilities of No Discrimination Based On			
			Time until		Number of	
Stimulus	vs Stimulus	Frogs Tested	First Call	First Chuck	Calls	Chucks
Whine-chuck[a]	chuck only	20	.032	.078	.007	.023
Whine-chuck	chuck-whine	10	.674	.465	.407	.317
Whine[a]	reversed whine	10	.013	.013	.007	.015
Whine-chuck[a]	reversed whine	10	—[b]	.142	.029	.032

Notes: Male descrimination among stimuli was tested by comparing four parameters of a male's response between each stimulus of a stimulus pair with a Wilcoxon matched-pairs signed-ranks test. The number of males tested is not always the sample size in each test, since tied pairs were omitted from the calculations.

[a] The stimulus that better elicited a vocal response (i.e., more responses and a shorter latency to response).

[b] Sample size was too small for statistical analysis due to the number of ties (from Ryan 1983a).

Table 5.13 Female *Physalaemus pustulosus* Response to Various Stimuli in Paired-Choice Experiments

Stimulus	# of responses	vs.	Stimulus	# of responses	P
Whine-chuck	5		chuck-whine	5	.6230
Whine	9		reversed whine	1	.0214
Whine	9		500 Hz tone	1	.0214
Whine	9		700 Hz tone	1	.0214
Whine	12		900 Hz tone	3	.0768

Source: Ryan 1983a.

I also tested the preference of females for a whine alone versus a constant pure tone within the frequency range of the whine, and of the same duration. Females were attracted preferentially to a whine when it was paired with either a 500 Hz or a 700 Hz pure tone (table 5.13). More females also approached a whine than the 900 Hz tone, but this difference was not statistically significant (table 5.13).

These results demonstrate that different components of the *P. pustulosus* complex advertisement call perform different functions. The chuck alone will not elicit responses from males or females; it must be combined with a whine, as it is in nature, to be biologically meaningful. A whine by itself does elicit responses from both sexes, but calls with chucks elicit more complex calls from males and preferential phonotaxis from females. This suggests that the whine is a necessary and sufficient stimulus for species recognition in *P. pustulosus*. Chucks contain information about male body size that females may use when selecting mates from among conspecific males, and chucks also mediate vocal interactions among males. Therefore, the chuck seems to function only in intraspecific interactions. The facts that the chuck is not sufficient for species recognition, that it results in preferential attraction of females, and that it increases predation risk to the caller further suggest that this portion of the call has evolved as a result of sexual selection. This is the first example of an anuran in which the advertisement call has different components, with one component functioning at the species level and the other at the individual level.

Narins and Capranica (1978) have shown that the different notes of the "co qui" advertisement call of *Eleutherodactylus coqui* also perform different functions, but functions are differentiated by sex. The *co* is used in male interactions and the *qui* to attract females. In fact, they showed sexual dimorphism in the tuning of the auditory nerve that matched the spectral properties of the call note to which each sex was most responsive. Narins and Capranica suggested that the division of function by different call components might be a general phenomenon in frogs. This phenomenon should become even more apparent with detailed studies of tropical anurans that tend to have more complex communication systems than their Temperate Zone counterparts.

Frequency Sweep and Call Recognition

The behavioral response by túngara frogs to frequency sweep direction raises questions about the sensory basis of species recognition in *Physalaemus*, and about the functional significance of frequency sweeps in anuran advertisement calls. Neurophysiological studies of the anuran auditory system have investigated the sensory basis of species recognition. Capranica and his colleagues (see reviews in Capranica 1976a, 1976b, 1977) have shown that the peripheral auditory system acts as a frequency filter, which in the majority of species studied tends to be tuned to those frequencies that characterize the species' advertisement call (see the brief discussion of the anuran peripheral auditory system in chap. 1). Although less is known about acoustical processing in the central nervous system, cells in the thalamus (Mudry, Constantine-Paton, and Capranica 1977), the telencephalon (Mudry and Capranica 1980), and the torus semicircularis (Walkowiak 1980; Fuzessery and Feng 1981; Narins 1983) respond to stimuli in a qualitatively different manner than do units in the periphery, thus suggesting a convergence of information from the peripheral auditory system in the central nervous system.

Sensitivity to sweep direction suggests that simple frequency filtering at the periphery, and enhanced response in the central nervous system to a combination of tones that maximally excite the basilar and amphibian papillae, is not sufficient to account for species recognition in *P. pustulosus*. A model for the decoding of frequency sweeps in bats has been suggested (Suga and Schlegel 1973; Suga 1978). Neurons in the central nervous system (specifically, in the inferior colliculus) of some bats are sensitive to sweep direction because they have a rejection mode—that is, an individual neuron not only has a frequency by which it is best excited, it also has a frequency (or a range of frequencies) by which it is inhibited. If a frequency sweep passes through the inhibitory frequency range before the excitatory frequency range the neuron will not fire. If, however, the sweep first passes through the excitatory frequency and then through the inhibitory frequency, the neuron will fire. Neurons that have their rejection mode at a frequency higher than their best excitatory frequency will be excited by an upward frequency sweep but inhibited by a downward sweep, even though the two sweeps may cover the same frequency range. Conversely, neurons inhibited by a frequency lower than their best excitatory frequency will be excited by a downward frequency sweep.

In the anuran peripheral auditory system, suppression of an excitatory response by the addition of a second, simultaneous frequency occurs and is called two-tone suppression (Capranica and Moffat 1980). This might suggest a mechanism for detection of frequency sweep direction in the periphery, but it seems unlikely for several reasons. Only those fibers from the amphibian papilla that are tuned to low frequencies, less than 500 Hz, are known to exhibit two-tone suppression (Capranica and Moffat 1980). Given the spectral characteristics of most frog calls, these fibers are unlikely to play any role in species recognition. Also, the inhibitory tone is always at a higher frequency than the best excitatory frequency of the unit. This would result in an inhibitory response to a downward frequency sweep according to the above model, which is not the case in the túngara frog.

Walkowiak (1980), and Fuzessery and Feng (1981) have shown that a large proportion of units in the torus semicircularis of fire-bellied toads, grass frogs (Walkowiak), and leopard frogs (Fuzessery and Feng) exhibit two-tone suppression. Interestingly, some of these units are inhibited by frequencies lower than the units' best excitatory frequency. This makes these units qualitatively different than those in the periphery that exhibit two-tone suppression, and suggests that their response properties are the result of neural convergence.

Capranica (1966) suggested that an anuran "mating call detector" in the central nervous system should show an enhanced response to a combination of tones that coincide with the frequency peaks in the species' mating call (perhaps as in responses found in the thalamus of the bullfrog by Mudry and Capranica 1980), and also should be responsive to species-specific temporal patterns of the call (Capranica and Moffat 1983). This study suggests that in some species a "mating call detector" also may have to detect features of the frequency sweep. At least among species in the genus *Physalaemus*, it appears that features much more subtle than only the direction of the sweep alone (e.g., the slope; see fig. 5.27) might also be decoded for species recognition among congenerics.

Calls with large frequency sweeps, by definition, encompass a broad range of frequencies. Such calls have at least two potential advantages: they might maximize locational cues, and they might provide more information in the frequency domain for use in species recognition. Whether some frogs evolved calls with extreme frequency sweeps because these calls confer some selective advantage is not clear. However, a possible advantage of a frequency sweep is that it might allow a male to broadcast a range of frequencies, in sequence, over a greater distance than could be traversed by a call in which the same range of frequencies is produced simultaneously. This is because the distance that any given frequency is transmitted is influenced by its intensity at the source. Given that both calls have the same total energy, in a call with a narrow frequency range, the energy in each frequency is greater than in the same frequency of a call with a broad frequency range. A narrowly tuned call will broadcast farther, but a broadly tuned call will contain a greater range of frequencies. A frequency sweep might result from selection to maximize both transmission distance and the range of frequencies transmitted. By concentrating its energy in a narrower frequency band and then modulating that frequency over time, a frequency sweep will transmit farther than a broadly tuned call but may still contain the same range of frequencies. Wiley and Richards (1982) make the same point in regard to bird song. Straughn and Heyer (1976) suggested that leptodactylid frogs that attract females to a breeding site from a long distance use calls with frequency sweeps, while those frogs in which males and females both reside near the breeding site use broadly tuned calls. This hypothesis is an interesting one that should be tested in the field.

Another, and not mutually exclusive, hypothesis is that reliance on features of frequency sweeps for species recognition allows finer subdivision of the acoustic environment. Species whose calls have extreme frequency sweeps appear to be more common in the tropics. Frog species diversity and the complexity of the nocturnal bioacoustic environment are greater in the tropics than in most regions of the Temperate Zone, and thus more information might be needed for species recognition in the tropics. Therefore, frequency sweeps can be added to the suite of spectral and temporal call features that encode species identity. In this context it is interesting that *P. pustulosus* and the South American bullfrog, *Leptodactylus pentadactylus*, sometimes utilize the same breeding site. The bullfrog's call covers a frequency range similar to that of the túngara frog, but the calls of these species sweep in opposite directions. A female túngara frog is not likely to mistake a bullfrog for a mate at a close distance, since the latter is so much larger (150 versus 30 millimeters; fig. 5.22). But bullfrogs eat túngara frogs, and a mistake by a túngara frog therefore would have far more serious consequences than the production of sterile hybrids.

Relative Attenuation of the Whine and Chuck

Not only do the whine and chuck differ in function, they differ radically in structure as well. These structural differences will be examined in the context of some theories that make predictions about the structure of acoustic signals.

In an acoustically ideal environment the intensity of a sound attenuates over distance, due to spherical spreading of the sound wave front at a rate that can be predicted by the inverse square law—a 6 dB loss of sound pressure for every doubling of distance traveled (Beranek 1954). The real world is far from acoustically ideal, and as a consequence sounds attenuate differentially depending on a variety of factors. Morton (1975) suggested that the habitat influences the evolution of bird songs in such a way that songs will be structured to decrease excess attenuation (i.e., attenuation in excess of that due to spherical spreading), or the corollary, that songs will be designed to maximize transmission distance. Morton compared the structure of bird songs in different habitats with the acoustic nature of the habitat, and he determined which frequencies were more severely attenuated. He showed that at elevations near or on the ground in forest habitats, song structure was matched to the local environment so that transmission distance was maximized. Although Morton's conclusion that different habitats attenuate frequencies in qualitatively different manners has been questioned or modified (e.g., Marten and Marler 1977; Marten, Quine, and Marler 1977; Wiley and Richards 1978; Ryan and Brenowitz, 1984; but see Bowman 1979), Morton's study was important in emphasizing the potential role of the environment in the evolution of animal communication. A number of subsequent studies have supported his general hypothesis that selection to maximize transmission distance might have some influence on the evolution of signal structure (e.g., Waser and Waser 1977; Bowman 1979; Gish and Morton 1981; Ryan and Brenowitz 1984).

Figure 5.28 Oscillograms of a whine plus chuck recorded at a near and far distance from the speaker.

Wiley and Richards (1978) pointed out that for many animal vocalizations, selection should not act simply to maximize transmission distance. They specifically suggested that signals that encode species recognition and individual recognition might be expected to attenuate at different rates. This hypothesis has been demonstrated in various groups of animals. Waser and Waser (1977) found that in 4 species of forest primates, long-distance calls possessed "design features" that reduced their attenuation rates relative to intragroup calls. This type of specialization also occurs in bird song. Brenowitz (1982) showed that the portion of the red-winged blackbird song that is necessary and sufficient to elicit species-specific behavior is also the portion of the song that attenuates less rapidly with distance, and this appears to result from the spectral structure of this portion of the song. Similar results were reported by Richards (1981) for the rufous-sided towhee.

The túngara frog call has components that function in either species recognition (the whine) or in intraspecific discriminations (the chuck); I tested the hypothesis that these call components attenuate differentially. I specifically predicted that the call component with the species recognition information—the whine—should attenuate less rapidly than the chuck. This prediction should be even stronger than that of other studies that predict differential attenuation of signals. Since a female is in close proximity to males before she selects a mate, the chuck probably does not function at a long distance, and some risk is involved in transmitting the chuck because of increased predation; the predation risk to the caller should increase with the distance over which the chuck is transmitted. Therefore, selection should result in rapid attenuation of the chuck.

To test this hypothesis, I broadcast a tape loop of a whine-plus-chuck call at ground level, and recorded the call at ground level at distances of 0.5, 1, 5, 10, and 20 meters from the speaker. The intensity of the call was 90 dB SPL at 0.5 meters from the speaker, approximately the intensity of a calling male. I determined the differential attenuation of the whine and chuck by comparing the difference in peak amplitude (in volts) between the whine and chuck at each distance (fig. 5.28). These experiments were conducted at three sites, two of which were typical *P. pustulosus* breeding sites and one which was not. Two sites were tested in 1980 and one in 1981. The habitat structure of these three sites differed considerably. The Gamboa site was a large flooded field of tall grass which had a large breeding population of túngara frogs during certain times of the year. The Chiva-Chiva site was an open gravel pit with no vegetation; túngara frogs also breed there. The site that was not a typical breeding habitat was in dense underbrush off of Snyder-Molino trail on Barro Colorado Island.

Sound attenuates most rapidly when the source and the receiver are at ground level (Michelson 1978)—the level on which communication occurs between the sexes for túngara frogs. The whine and chuck attenuated so rapidly that only at one site, Chiva-Chiva, was I able to distinguish the waveform of the call from the background noise at all distances tested (table 5.14). However, when I compared the differences in the relative amplitudes of the whine and the chuck at different distances from the speaker, the trend was the same for all sites. The differences in the peak amplitude between the whine and the chuck is greatest at 0.5 meters from the speaker, with the chuck having the greater peak amplitude at all sites (table 5.14). As the distance between the microphone and the speaker increases, the differences in peak amplitude between the whine and the chuck decrease. At 20 meters at the Chiva-Chiva site almost no difference in peak amplitude exists between the two call components, and at 5 meters at the Gamboa site the peak amplitude of the whine is actually slightly greater than that of the chuck (table 5.14). These data demonstrate that at the sites tested, the chuck does attenuate more rapidly than the whine. We cannot conclude from these results, however, that selection for differential attenuation of call components has occurred. We will consider alternative explanations for this phenomenon shortly.

Table 5.14 Mean Difference between Peak Amplitudes of the Whine and Chuck

	Snyder-Molino	*Chiva-Chiva*	*Gamboa*
0.5 meters	− 1.31 (0.2)	− 1.29 (0.4)	− 1.65 (0.5)
1.0 meters	0.33 (0.3)	− 1.16 (0.4)	− 0.94 (0.3)
5.0 meters	ND	− 1.03 (0.1)	0.13 (0.01)
10.0 meters	ND	− 0.19 (0.03)	ND
20.0 meters	ND	− 0.02 (0.02)	ND

Notes: Standard deviation given in parentheses. Mean difference given in volts. ND indicates that the signal could not be distinguished from the background noise.

Frequencies attenuate differentially in most environments (Embleton, Piercy, and Olsen 1976). In general, higher frequencies tend to attenuate more rapidly than lower frequencies. But for sound sources at low heights, ground absorption and interference between direct and reflected sound waves is more severe, and some lower frequencies especially do not transmit well. This sometimes results in a "frequency window," from about 1 kHz to 4kHz,

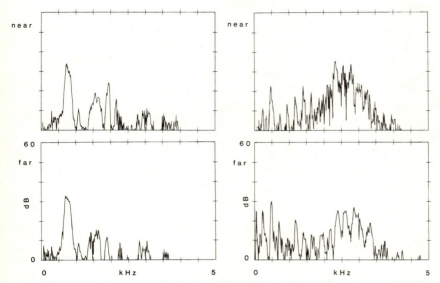

Figure 5.29 A power spectrum of the first 80 milliseconds of a whine recorded at a near and far distance from a speaker.

Figure 5.30 A power spectrum of a chuck recorded at a near and far distance from a speaker.

in which frequencies transmit best (Morton 1975; Marten and Marler 1977; Marten, Quine, and Marler 1977). Still, whether frequency attenuation differs among habitats is not clear (Marten, Quine, and Marler 1977; Wiley and Richards 1978; Michelson 1978; but see Bowman 1979).

The reason why the whine attenuates less rapidly than the chuck is revealed by an examination of the changes in the power spectrum (i.e., the relative amplitude-frequency profile) with distance for each of the call components. I compared the power spectrum for each call component at 0.5 meters from the speaker with the power spectrum at the farthest distance available for each site. The power spectrum shows the relative concentration of energy at different frequencies. For example, in the initial part of the whine, peak energy concentration is about 750 Hz (fig. 5.29), and the chuck has most of its energy in harmonics between about 1500 Hz and 2500 Hz (fig. 5.30). By comparing the power spectrum of a call component between distances, the relative attenuation of each frequency can be determined.

For the power spectrum of the initial 80 milliseconds of the whine recorded at 0.5 meters, I arbitrarily set the dB level of the dominant frequency, 747 Hz, to 0 dB for each distance. Since the dominant frequency, by definition, had the most energy, all of the other frequencies had negative dB values. I then determined the relative attenuation of a given frequency by subtracting the dB level for that frequency at the far distance from the dB level for the same frequency at the close distance. If the resulting difference in dB level was negative, this indicated that the frequency being examined attenuated more rapidly than the dominant frequency. A positive difference indicated that it attenuated less rapidly. This result allowed me to compare the relative attenuation of harmonics within the whine.

At two of the three sites (Snyder-Molino and Gamboa), the dominant harmonic (747 Hz) attenuated less rapidly than all (Gamboa) or all but one (Snyder-Molino) of the other harmonics in the whine. At the Chiva-Chiva site, all the other harmonics attenuated less rapidly than the dominant over the 20 meters, but the difference was less than 3 dB for 6 of the 8 harmonics (fig. 5.31). The energy in the first part of the whine is concentrated at 747 Hz and energy decreases in the upper harmonics (fig. 5.29). These results indicate that energy in the whine is concentrated in the harmonic that attenuates least in the habitats tested.

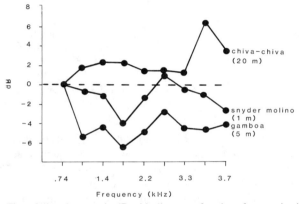

Figure 5.31 The relative decrease in dB with distance of various frequencies in the first 80 milliseconds of the whine.

The same analysis was conducted with the chuck, with the dB level of the chuck's dominant frequency, 2.3 kHz, set to 0 dB. If we examine the frequency band that contains most of the chuck's energy, 1500 Hz to 2500 Hz, it is seen that these frequencies attenuated more rapidly than the lower frequency harmonics (200 Hz to 1500 Hz; fig. 5.32). At the Gamboa site, the dominant frequency band also attenuated more rapidly than the higher frequency band (2500 Hz to 4000 Hz). At Chiva-Chiva there was the same trend but to a lesser degree, and at the Snyder-Molino site the midfrequency band transmitted better than the higher frequency band (fig. 5.32). In contrast to the whine, the energy in the chuck is not concentrated in those frequencies that attenuate least, and in some cases the dominant frequency band is the least appropriate band for energy to be concentrated in order to maximize transmission distance.

These results demonstrate that the call component containing the species recognition information—the whine—attenuates less over distance than the other call component—the chuck. This is due primarily to the spectral

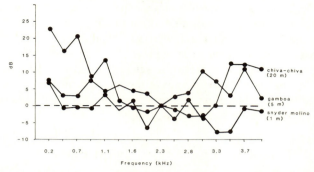

Figure 5.32 The relative decrease in dB with distance of various frequencies in the chuck.

structure of the call: the whine contains most of its energy in frequencies that transmit better than do the frequencies that contain most of the energy in the chuck. These results are consistent with predictions and results of studies dealing with the structure of signals used in long-range communication (e.g., Waser and Waser 1977; Richards 1981; Brenowitz 1982).

Whether the structure of the chuck evolved in response to selection for decreased transmission distance, or whether there are other reasons why the chuck has a structure that happens to attenuate more rapidly, is not clear. At least two alternatives exist to the hypothesis that the structure of the chuck has resulted from selection to increase its rate of attenuation. First, a general consistency was noted by Darwin (1872) between the structure of an acoustic signal and the perceived (by the observer) motivational state of the signaler. For example, an animal in a state of appeasement might produce tonal (pure frequency) sounds while the same animal in an aggressive state might produce calls that are noisy or rich in harmonics (Darwin 1872; Collias 1960; Morton 1977; Maier 1982). Morton (1977) elaborated this concept, and according to his "motivation structural rules," he specifically suggested that during encounters in which it would be advantageous for a male to signal large size, selection for low frequency sounds should occur, and these low frequency sounds probably would be noisy or rich in harmonics. Since the chuck does provide some information about male size, and females seem to use this information when selecting mates, the spectral structure of the chuck is consistent with the hypothesis proposed by Morton. Another explanation for the structure of the chuck was discussed earlier in this chapter. Marler (1955) suggested a convergence in structure of signals that function to reveal the location of the signaler. For animals that locate sounds by making binaural comparisons, a signal with rapid rise and decay times and a rich harmonic structure should be easier to locate than signals lacking these characteristics. I have extensively discussed both the possibility that the chuck increases localizability, as well as the lack of data to support this theory.

The fact that the chuck is structured in a manner that concentrates energy in certain frequencies explains why it attenuates more rapidly. Also, a male túngara frog producing a chuck that attenuates more rapidly with distance would, in the long run, have a lower risk of predation from acoustically foraging predators than would a male producing a chuck that travels farther. However, it is difficult to discern the selective forces, if any, that are responsible for the evolutionary origin of the structure of the chuck. As with much of the theory in behavioral ecology, eliminating all but one of the many competing hypotheses is not always possible.

Summary of the Role of the Advertisement Call

The complexity of the túngara frog advertisement call varies. All calls consist of a whine followed by 0 to 6 chucks. The whine is about 400 milliseconds long; its dominant frequency begins at about 900 Hz and falls to about 400 Hz. The whine also contains substantial energy in the second to fourth harmonics. The chuck temporally overlaps the whine, is of much shorter duration, and has energy in all of the first thirteen harmonics of the fundamental frequency of 200 Hz to 250 Hz. Drewry, Heyer, and Rand (1982) demonstrated that two separate but partially coupled structures are responsible for the whine and the chuck. Changing tension of the vocal cords results in the whine, and the chuck is produced by changes in larynx shape that introduce a pair of fibrous masses into the air flow from the lungs.

When males call in isolation, they usually produce a whine only, but they add chucks to their calls in response to the movement of nearby frogs and to the vocalizations of other conspecific males. The more complex a stimulus call (i.e., containing more chucks), the more complex is the male's response. When given a choice between calls, females are attracted preferentially to complex calls. Thus the question is raised: Why do not males maximize their mate attraction ability by always producing the complex calls?

Although males obtain a benefit by producing complex calls, in that they are more likely to attract mates, some disadvantage to producing chucks must exist. An analysis of energy expenditure during calling shows that no more energy is required to produce calls with chucks than calls without chucks. However, there is a cost of increased predation risk. Males are more likely to be captured by the frog-eating bat, *Trachops cirrhosus*, when they are producing calls with chucks. Therefore, the túngara frog's call complexity series allows males to effect a compromise between maximizing their ability to attract mates and minimizing their risk of predation.

In the field, larger male *P. pustulosus* are more likely to acquire mates. A negative correlation occurs between male size and the fundamental frequency of the chuck. When given a choice, females usually prefer calls with lower frequency chucks, although when the frequency difference was only 30 Hz the females did not exhibit a preference. This preferential female phonotaxis appears to be partially responsible for the size-biased male mating success observed in the field.

Testing the hypothesis that female choice has influenced the evolution of the frequency structure of túngara frog calls is difficult. When compared to the calls of other members of the subfamily Leptodactylinae or only to other congenerics, the *P. pustulosus* call has a considerably lower frequency than would be expected given the size-frequency relationship of closely related species—more so than any of the other species considered. Outgroup comparisons suggest that the low frequency call is the derived state. The fundamental frequency of the chuck is responsible for the large departure of the túngara frog call from the general relationship. Properties of the larynx and the vocal sac result in the ability of *P. pustulosus* to produce and transmit these relatively low frequency calls. The results are consistent with the hypothesis that sexual selection has resulted in the relatively low frequency calls of *P. pustulosus*.

The chuck provides information used by the female in intraspecific mate choice: females are attracted preferentially to calls with chucks, and they prefer calls with lower frequencies. The chuck is not necessary to elicit phonotaxis from females or vocal responses from males. Again, this conclusion suggests that the chuck has evolved under the influence of sexual selection, particularly female choice. The whine, however, is both a sufficient and a necessary stimulus for evoking species-specific behaviors from both males and females. The direction of the frequency sweep of the whine partially encodes species identification—it must be in the proper direction to evoke preferential phonotaxis from females and maximum vocal responses from males. A review of the structure of calls of sympatric *Physalaemus* in Argentina suggests that differences in the form of the calls' frequency sweeps might be important for species recognition within this genus.

Theories of the structure of acoustic communication signals suggest that signals containing species or group recognition information should be designed to transmit over longer distances than signals used to communicate relevant information about individuals (e.g., body size). This should be even more important when comparing the attenuation rates of the whine and the chuck, because a disadvantage to males transmitting chucks over long distances should exist from increased predation risk from acoustically foraging predators. Experiments showed that at all of the three sites tested the chuck does attenuate more rapidly than the whine. These results are consistent with the above hypothesis on the structure of long-range acoustic signals, but they are also consistent with the hypotheses that motivation structural rules, or selection for locatability have resulted in the spectral structure of the chuck.

This portion of the study has reviewed the complexity, both in structure and function, of the advertisement call of the túngara frog. These results should show that no longer can this complex behavior only be considered to have the simple function of attracting conspecific females and repelling conspecific males. These results should also show that not only must the potential advantages of the call be considered (e.g., attracting females), but some of the costs involved in this behavior must be considered as well before a full understanding of its function can occur, and that a phylogenetic analysis is necessary to understand the patterns of evolutionary change. In the next chapter I consider possible explanations for why females are preferentially attracted to the calls of some conspecific males over others.

6 Why Do Females Choose Mates?

The question of why females choose mates can be addressed at two levels—the proximate and the ultimate. The previous chapter has addressed the proximate question; females choose larger males as mates because they are attracted preferentially to spectral characteristics of calls produced by larger males. In this chapter the question of female choice will be addressed at the ultimate level: What factors account for the evolutionary origin and maintenance of this female behavior?

Female mate choice usually results in females mating with conspecific males. This choice is of obvious importance to a female, since heterospecific genomes often are uncomplementary, and viable offspring do not result. Therefore, heterospecific matings can be a total waste of reproductive investment. This selective advantage of female mate choice will be assumed, and the following discussion of the evolution of female choice will be restricted to differential mate choice at the intraspecific level.

Female mate preference can evolve under the influence of sexual selection or natural selection. The female preference can be favored by natural selection if the choice of a male increases the number of offspring produced by a female (a "nongenetic" benefit), or if the offspring of the female acquire genotypes that are superior for survival (a "genetic" benefit). A female preference will evolve under the influence of sexual selection if the result of female choice is a genetic correlation of the male trait and the female preference. This will result in the coevolution of the trait and the preference. The sons of females exercising the preference will more likely be selected as mates by females, and the daughters will inherit the preference for mating with males possessing these traits. As males with the preferred trait increase in the population due to their increased mating success, the alleles for the female preference also increase because they are in linkage disequilibrium with the alleles for the male trait (see Arnold 1983). As was detailed in chapter 1, this process is called runaway sexual selection. A third possibility is that female mate preference could merely be an epiphenomenon of nonheritable variation in the male trait interacting with the sensory properties of the female's nervous system. In this instance, the female preference for conspecific mates need not evolve under the direct influence of either natural or sexual selection.

Female Mate Choice as a Neutrally Adaptive Trait

A possible example of female choice as a neurophysiological epiphenomenon comes from Capranica and his colleagues' work with bullfrogs, reviewed briefly in chapter 1 (see also Capranica 1976a). Large males have a lower frequency peak in their advertisement call between 200 Hz and 300 Hz. The frequency tuning of the peripheral auditory system of both sexes closely matches the spectral characteristics of the call, including the lower frequency peak. Smaller males have a lower frequency peak of 700 Hz to 800 Hz, and this peak shifts to 200 Hz to 300 Hz as the males grow larger. These data suggest that female bullfrogs should respond preferentially to the lower calls of larger males. This preference could be due only to the response properties of the female's auditory system, and the evolution of these properties could well be unaffected by sexual selection. For example, the close match between the female auditory system and the advertisement call of large males could result from selection for species recognition. (Bullfrogs might not be the best example since Howard 1978a and 1978b has documented a selective advantage to female choice of larger males. But the point still holds.)

Unfortunately, no data are available on the frequency tuning of the peripheral auditory system of *P. pustulosus*. Nevertheless, the possible constraints of the female's auditory system on her mate preference in this species is still worth considering. As was pointed out in chapter 1, the amphibian papilla of the frog's peripheral auditory system usually gives rise to two populations of nerve fibers that respond best to low frequencies (e.g., around 250 Hz in bullfrogs; Feng, Narins, and Capranica 1975) or midfrequencies (around 600 Hz in bullfrogs). Fibers from the basilar papilla are most responsive to higher frequencies (1400 Hz in bullfrogs). Among frog species, the basilar papilla tends to be tuned to higher frequencies in smaller species, for example, 3100 Hz to 3800 Hz in *Hyla cinerea* (Moffat and Capranica 1974) and 2000 Hz to 3700 Hz in *Eleutherodactylus coqui* (Narins and Capranica 1980). This suggests that the dominant frequency of the whine component of the túngara frog call (about 900 Hz to 400 Hz; fig. 5.2) does not excite the basilar papilla, although this papilla might be excited by some of the upper harmonics of the dominant frequency. However, females will respond to synthetic whine-only calls that contain only the dominant frequency sweep and no other harmonics, that is, no frequencies above 900 Hz. Species recognition by *P. pustulosus* might be accomplished by excitation of only the amphibian papilla.

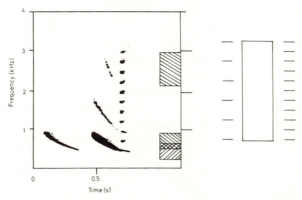

Figure 6.1 Túngara frog calls with zero and one chuck. The bars on the right represent the hypothetical frequency windows derived from three populations of nerve fibers innervating the inner ear. It is assumed that the two bars with the lower frequency distributions arise from the amphibian papilla, and the bar with the higher frequency distribution arises from the basilar papilla. The bar on the far right of the figure represents that range of frequencies to which the basilar papilla is sensitive. On each side of the bar is a hypothetical chuck from the túngara frog call. The chuck on the right has a fundamental frequency one-half that of the chuck on the left.

If the peripheral auditory system of *P. pustulosus* is similar to that of other frogs investigated, then the upper harmonics of the chuck do excite the basilar papilla (fig. 6.1). The frequency sensitivity of the basilar papilla of túngara frogs is not known, but three properties of the chuck must determine the extent to which it excites the basilar papilla. The first is frequency range: maximum excitation of the basilar papilla can occur only if the chuck covers the entire range of frequencies to which the basilar papilla is sensitive. The second is the relative intensity of the harmonics of the chuck. Given a fixed number of harmonics, the chuck with relatively more energy in the harmonics within the frequency range exciting the basilar papilla will elicit a greater neural response. The third property of the chuck that influences the amount of excitation of the basilar papillar is its fundamental frequency. As the fundamental frequency of the chuck decreases, the number of harmonics that fall within the excitatory frequency range of the basilar papilla increases, as long as the frequency range of the chuck encompasses the frequency range to which the basilar papilla is sensitive. For example, a chuck with a fundamental frequency one-half that of another chuck will, on average, have twice as many harmonics that excite the basilar papilla (fig. 6.1). This occurs because harmonics are integer multiples of the fundamental frequency. A fundamental frequency of 200 Hz has its first three harmonics at 200 Hz, 400 Hz, and 600 Hz, while a fundamental frequency of 250 Hz has these same harmonics at 250 Hz, 500 Hz, and 700 Hz.

In a more realistic example, if the basilar papilla were sensitive to frequencies from 1500 Hz to 2500 Hz, then a chuck with a fundamental frequency of 200 Hz would have five harmonics within that frequency range, while a chuck with a fundamental frequency of 260 Hz would only have four. Everything else being equal, the lower frequency call would excite the basilar papilla with 20% more frequencies. However, if two calls had the same total energy content, then chucks with fewer harmonics would have more energy in each harmonic. If a linear relationship exists between energy in a harmonic and the neural excitation it elicits, then the two calls might be equivalent. But if the fibers saturate below the maximum intensity of the harmonic, then the calls with more harmonics will elicit a greater neural response.

In the absence of the necessary neurophysiological data, anticipating the effect of the harmonic structure of the chuck on the excitatory response of the female's basilar papilla and on the mechanism of her preferential phonotaxis is not possible. Even if these results were available, another leap of faith would be needed to extrapolate these results to the evolution of female mate choice. However, this line of questioning makes an important point: as we consider factors responsible for the evolution of mate preference, we must keep in mind the possibility that this behavior could be due to response properties of the female's sensory system, and that these response properties have evolved for reasons totally unrelated to intraspecific mate choice. In the terminology of Williams (1966) then, the observed behavior of female choice among conspecific males could be an effect and not an evolved function. Of course regardless of why the female preference exists, it still can influence the evolution of the male trait.

Potential Genetic Effects of Female Choice

Ectothermic vertebrates continue to grow after reaching sexual maturity. Thus large males have either survived longer or grown faster. Trivers (1976), among others, has suggested that females prefer to mate with larger males because these males have maximized some combination of growth and survivorship, and that the female preference evolves because this natural selection advantage is passed on to the female's offspring.

Can this scenario be applied to túngara frogs? Not surprisingly, the data needed to test the hypothesis of eugenic mate choice are not available for *P. pustulosus*, nor for almost any other species. However, some of the problems in applying this hypothesis in general, and to these frogs in particular, are worth exploring. First, we must ask why some males are larger than others. Since ectotherms often continue to grow after reaching sexual maturity, larger males are usually older than smaller males. But in *P. pustulosus*, as well as in some other frogs (e.g., bullfrogs; Howard 1981), growth rates decline with male size ($r_s = -.90$, $P < .01$; fig. 6.2). This decline tends to decrease the correlation between male size and age. A more serious problem is that large variations exist in growth rates among individuals (fig. 6.2). Given this variation, body size is even less likely to indicate age

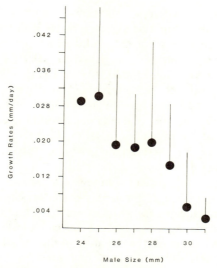

Figure 6.2 The average growth rates of male *P. pustulosus* of different sizes (snout-to-vent length, SVL) plus one standard deviation.

reliably. Although, of the possible phenotypic indicators of age that are available, size probably is the best.

One could argue that age is not an important criterion, that when they select larger males, females acquire mates that have maximized some combination of growth and survivorship, as suggested by Trivers (1976). A male túngara frog will be larger if he lives longer or grows faster; however, males that rarely attempt to breed should be large for their age as well since they do not expend energy used in sexual displays, and they probably live longer because they avoid predation at the breeding site. Some larger males could actually mate less frequently than males that die sooner, and smaller, but spend more time at the breeding site. If this were so, females would be selecting for a genotype that leaves fewer offspring. This type of preference would be unlikely to increase in the population.

At least one other possible mechanism of mate choice based on a genetic, natural selection advantage must be considered. Optimality theory has been an assumption, either implicit or explicit, of most discussions of this type of female mate choice. These discussions all have considered the possibility that females choose mates with the "best" alleles. However, the female mating strategy could have evolved under the influence of a more conservative selective advantage. By avoiding the smallest males, females might be less likely to mate with males that have certain genetic deficiencies, such as poor resistance to diseases and parasites (as suggested by Hamilton and Zuk 1982), metabolic deficiencies, neuromuscular problems that result in decreased foraging ability, and a host of other possibilities. By choosing larger males, female *P. pustulosus* might not be selecting the males with the "best" alleles, but they might be avoiding males with the "worst" alleles.

The theoretical arguments for the evolution and maintenance of female choice by a genetic advantage through natural selection are controversial (e.g., Zahavi 1975, Anderrson 1982b, Arnold 1983, O'Donald 1983). Certainly this problem will not soon be resolved empirically, although detailed genetic models may shed some light on the internal consistency of the theory. Thus while not very satisfying, the only conclusion from this section is that adequate data do not currently exist to determine if female *P. pustulosus*, or almost any other species, increase the genetic quality of their offspring in relation to natural selection.

Another point must be considered in regard to a genetic natural selection advantage to female choice. If a trait existed that was correlated with some natural selection advantage, and females preferred that trait, then the trait would also confer a mating advantage and thus be subject to sexual selection. This is how Fisher envisioned the initiation of the sexual selection process. The direction of natural selection and sexual selection can be the same, but if they are not the same then we expect to see a male trait that represents some compromise between the two forces. The female preference exhibited, therefore, is not a product only of natural selection.

Given the theoretical problems in applying theories of a genetic natural selection advantage, we can contrast predictions made by the natural selection and sexual selection theories in regard to the evolution of male traits in túngara frogs. The natural selection theory predicts selection for increased growth and survivorship, which should be evidenced by evolution of large body size. This theory does not necessarily make any prediction about how the frequency of the call should evolve. The sexual selection theory predicts that female choice should result in the evolution of lower frequency calls. This hypothesis does not make any prediction about how body size should evolve. Therefore, both theories allow for the joint evolution of large body size and low frequency calls, which in general are correlated characters. If túngara frogs were very large and had extremely low frequency calls, the two hypotheses could not be discriminated. If they were very small with high frequency calls, then both hypotheses would be rejected. However, as the evidence in the previous chapter shows, body size and call frequency have become uncoupled to some extent in túngara frogs. Relative to closely related species, the túngara frog's call is much lower in frequency, but its body size is within the range of the other species (figs. 5.21, 5.23; table 5.9). Thus the evolution of male traits is consistent with the notion that the traits have evolved under sexual selection and not under natural selection for increased growth and survivorship.

Before leaving this section I want to briefly discuss the problem of heritability of male fitness traits. I will immediately state my bias—to some extent I think the "problem" is a red herring.

Williams (1975) suggested that female choice of traits associated with male fitness would result in the depletion of additive genetic variation for those traits. Therefore, heritability would be reduced drastically, and thus evolution would become impossible. This argument has become very popular and has led many authors (e.g., Searcy 1982) to reject the hypothesis that female choice influences the genetic constituency of the female's offspring. Should this line of reasoning lead us to reject genetic hypotheses that seek to explain the evolution of male secondary sexual characteristics and female preference for these characteristics?

Darwin (1871) documented the extensive diversity of secondary sexual characteristics in the animal kingdom. Different species have different characteristics, and even in the absence of selection theory most everyone would agree that these differences have evolved. If they evolved, of course, heritable genetic differences had to exist in the traits among males. Therefore, the reasoning proposed by Williams, even if true in the most extreme form, cannot be used to argue against theories explaining the evolutionary origin of these traits. This logic is obvious to the point of being trivial, and surely Williams did not argue against the role of female choice in the origin of male secondary sexual characteristics. But there have been some extreme interpretations of Williams's argument.

The real controversy, however, centers on whether sufficient heritability of male traits exists to maintain female mating preferences. Two considerations must be examined here. The first is, To what extent is additive genetic variation replenished as it is being depleted through female choice? Lande (1976, 1981) contends that normal mutation rates at loci of polygenic traits maintain sufficient genetic variation on which selection can act. He also believes that the problem is not a problem. A rather unique suggestion has been made as to another mechanism that might maintain variation of male traits selected by females. Hamilton and Zuk (1982) proposed that heritability for genetic resistance to pathogens and parasites should be maintained because of the coevolutionary cycle of host resistance and parasite virulence. A male's resistance to various parasites should be reflected in some phenotypic indication of health and vigor, and females selecting males on the basis of these phenotypic traits should be selecting males of superior phenotypes. This theory still needs further testing, and it must be evaluated in terms of the interaction between natural selection and sexual selection. Data exist that address this topic. As pointed out in chapter 5, evidence can be found for heritable variation for male courtship traits that are under sexual selection by female choice (crickets, Cade 1981; fruit flies, MacDonald and Crossley 1982).

The second aspect of the lack-of-heritability argument that needs to be addressed is, How necessary is additive genetic variance for the maintenance of a female preference? Fisher (1958), West Eberhard (1979, 1983) and Lande (1980, 1981) all acknowledge that when genetic variation arises, a rapid evolutionary response of traits under selection should occur, and this response should continue until additive genetic variance is depleted. This hypothesis predicts that sexually selected traits should evolve in a punctuated manner, and suggests that periods with little or no heritability for male traits under selection will occur. If this is so, can the female preference be maintained during periods of genetic stasis?

West Eberhard (1979) points out that female choice is always based on the phenotype, and that selection (but no evolutionary response to selection) will continue in the absence of a genetic component to the phenotypic variation. She suggests that most often female choice is based on nonheritable variation. However, when genetic variation arises, the trait evolves.

Anderrson (1982a) presents evidence suggesting that female preference can be maintained during periods when there is no significant phenotypic variability in the trait among males. Male long-tailed widow birds—a species of polygynous weaver bird—have tails that are about 0.5 meters long. Although sexual selection theory might predict that the long tails are the result of female choice, Anderrson found that no correlation exists between male tail length and the number of nests on a male's territory. Also not much variation in tail length was found among males. However, when he experimentally lengthened or shortened a male's tail, he discovered that tail length did influence a male's reproductive success; males whose tails were experimentally lengthened had more active nests on their territories. This finding supports the hypothesis that long tails of widow birds are a product of sexual selection by female choice, even though under normal circumstances this preference was not exhibited, presumably because not enough variation existed on which females could base their choice. These data also support West Eberhard's (1979) suggestion that female choice will be "ready" to operate on heritable components of fitness as they arise.

The discussion presented by West Eberhard (1979) suggests that the current state of a sexually selected trait may reflect a historical response to sexual selection, even though heritable variation of the trait may or may not be seen at present, and that one could still expect to see female choice in the absence of heritability of male fitness traits. Depending on the periodicity of the evolutionary responses of a trait to sexual selection, a genetic effect of female choice might be difficult to demonstrate experimentally. The work of Partridge (1980), demonstrating competitive superiority of larvae of females who chose mates relative to females who did not, is one of the very few exceptions.

Clearly the potential genetic effects of female choice is a very controversial topic. Most of the discussions of female choice as a potential component of a runaway sexual selection process assume that no selection on the female preference occurs. But recently Parker (1982, 1983) has suggested that in some species considerable cost might be incurred by females choosing mates, and that this cost should affect the evolution of female choice. This

Nongenetic Effects of Female Choice

Although selection is necessary for females to avoid heterospecific matings, Fisher (1958) discusses how this selection can result in female preference among conspecifics. Consider a trait that identifies a species and encompasses a certain range of variation. If at one extreme of the variation this trait is similar to a species-identifying trait of a heterospecific, then female choice of males with the trait at the opposite extreme would increase her probability of mating with a conspecific, and selection for this preference should occur. With the exception of the possibility that her sons would inherit the male's trait, conspecific males would not provide differential benefits to females, but this female preference would be maintained by natural selection because of the deleterious consequences that heterospecific matings can have on a female's fertility. This type of interaction can eventually lead to character displacement of species-identifying traits. For example, *Hyla ewingi* and *H. verreauxi* have similar advertisement calls when allopatric, but in sympatry their calls are more different (Littlejohn 1965). Gerhardt (1982) has suggested that interspecific interactions might influence intraspecific mate choice in some North American hylid frogs, although Paterson (1982) has questioned the validity of the reproductive character displacement theory. Clearly, this is not the case in *P. pustulosus*, since intraspecific mate choice is based on a call component that does not play any role in species recognition.

That females can gain nongenetic benefits through intraspecific mate choice is well documented (e.g., see reviews by Thornhill 1980; Searcy 1982) and indicates that a female preference can potentially evolve under the influence of natural selection. The best documented case for anurans is Howard's (1978a, 1978b) study of bullfrogs. Male bullfrogs defend territories and females deposit their eggs on the territories of their mates. Females mate preferentially with larger males on better territories. These territories are "better" in the sense that eggs deposited there have greater hatching success because they suffer less predation by leeches and fewer developmental abnormalities resulting from extreme temperatures. Whether female choice is based on the male or the territory is not clear, but choice based on either would yield the same results. Howard suggests that females might also gain genetic benefits from mating with larger males, since these males are older and thus have demonstrated that their genotypes are adapted for survival.

Male túngara frogs have few opportunities to provide nongenetic benefits to females. Males do not defend territories and do not compete for oviposition sites. Thus differential control of resources by males cannot influence mate choice by female *P. pustulosus*. The only benefits provided by males are sperm and construction of the foam nest. Although some variation in the size and shape of the nest has been noted, this variation is not related to male size. No data exist indicating that nest size and shape influence the adaptive value of the nest (e.g., protection from predators, resistance to rainstorms), although it would not be surprising if such a subtle influence were discovered.

The remaining benefit is sperm, or more specifically, a male's ability to fertilize the eggs. Although fertilization is usually considered an efficient process (e.g., Kluge 1981), in a number of species males do not fertilize all of the females' eggs (Davies and Halliday 1977; Smith-Gill and Berven 1980; Kluge 1981; Kruse and Mounce 1982). Licht (1976) presents data demonstrating that toads (*Bufo americanus*) mate assortatively; a positive correlation exists between the sizes of mated males and females. He suggests that when mates are of the appropriate relative sizes, the rate of fertilization will be maximized, but he does not present data to support this contention. Davies and Halliday (1977) showed that *B. bufo* also mate assortatively with respect to size, and they demonstrated that more eggs were fertilized in the laboratory when males and females were closer in size. In the studies by Licht, and Davies and Halliday, the role of female choice in determining the ultimate pairing of mates is not clear (see chap. 1), nor is it known if fertilization success in *B. bufo* is as low in the field as it was in the laboratory.

I tested the hypothesis that in túngara frogs, the size of males also influences fertilization rate. It was necessary to test this hypothesis under controlled conditions. The eggs of *P. pustulosus*, unlike the transparent egg mass of bullfrogs (Howard 1978b), are not visible when they are in the foam nest; therefore, monitoring fertilization success in the field is not possible. Another problem is that pairs often nest communally (fig. 3.14), so that determining the parents of offspring from all nests is not possible. Therefore, I collected males and females in the field and allowed them to mate in the laboratory. Each pair nested in a separate plastic bucket. After three to four days I counted the total number of eggs that hatched (i.e., the number of tadpoles), and then dissolved the nest and determined how many eggs did not hatch. Unhatched eggs showed no signs of development, so I assumed they were not fertilized.

This procedure was conducted for 68 nesting pairs. For each pair, the male's snout-to-vent length, the female's snout-to-vent length, the number of hatched eggs, and the number of unhatched eggs were recorded. The mean clutch size was 234.2 (standard deviation = 97.6). As with many species of frogs (Salthe and Duellman 1973), a significant correlation was noted between female size and clutch size ($r = .54$, $P < .01$; fig. 6.3). Because this species produces many clutches during the season and deposits them opportunistically during favorable environmental conditions (see chap. 3), the fact

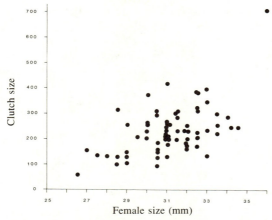

Figure 6.3 The number of eggs per clutch for female *P. pustulosus* of different sizes.

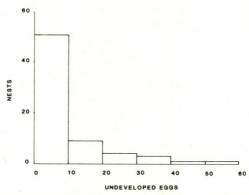

Figure 6.4 The frequency distribution of the number of undeveloped eggs per nest for *P. pustulosus* (from Ryan 1983b).

that female size explains only 29% of the variation in clutch size was not surprising.

Most of the eggs produced in this experiment hatched successfully. However, some nests had a large number of eggs that remained undeveloped (fig. 6.4). The average number of unhatched eggs per clutch was 9.1. This means that on average, 4% of the energy invested by females in reproduction (see chap. 7) is wasted. The percentage ranged from 0% to 76%.

Although large clutches, by definition, contained more eggs, large clutches were no more likely to have unhatched eggs than were smaller clutches (table 6.1). Since female size and clutch size are correlated, it is not surprising that female size was not correlated with the number of unhatched eggs (table 6.1). If eggs did not hatch because they were not fertilized, which appeared to be the case, we would expect the number of unhatched eggs to increase with clutch size if the probability of not being fertilized was the same for all eggs in all clutches. These results suggest that failure of fertilization is not a totally random event.

Table 6.1 Spearman Rank Correlations between Various Parameters of Mate Size and Reproductive Output

	Clutch Size	Unhatched Eggs	Absolute Value of Male-Female Size
Male size	−.223	.115	−.589***
Female size	−.538***	.079	.435**
Clutch size		.056	.294*
Unhatched eggs			.300*

* $P < .05$.
** $P < .01$.
*** $P < .001$.

Females preferentially mate with larger males in the field, and on average females suffer a 4% loss of reproductive output due to unhatched eggs. Could these two phenomena be related? I tested the hypothesis that by selecting larger males, females increase their reproductive output because more eggs are fertilized. The data do not support this hypothesis: no significant relationship exists between male size and the number of unhatched eggs in the clutch (table 6.1). However, when I examined the data more closely, the size of a male relative to his mate did seem to influence the number of unhatched eggs. In fact, a significant correlation existed between the absolute value of the size difference between mates and the number of unhatched eggs (table 6.1; fig. 6.5). Females usually are larger than males (fig. 6.6); therefore, male and female size also were correlated with the size difference between mates in the expected directions (table 6.1). And clutch size is correlated with female size, so it too was correlated with mate size difference (table 6.1).

The only variable that statistically was significantly related to the number of unhatched eggs was the size difference between mates. However, it was desirable to analyze the influence of the interrelated variables on the number of unhatched eggs. Using a multiple regression analysis, the effect of male

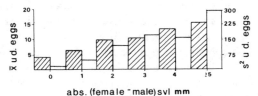

Figure 6.5 The mean (striped bars) and variance (open bars) of the number of unhatched eggs per nest as a function of the absolute size difference between mates (snout-to-vent length, SVL; from Ryan 1983b).

size, female size, and clutch size on the number of unhatched eggs was considered before and after the variation explained by the size difference between mates was removed. Because the number of unhatched eggs per nest was distributed in a Poisson-like manner (fig. 6.2), these data were transformed by square root prior to analysis (Steele and Torrie 1960).

The combined variables explain a significant amount of the variation in the number of unhatched eggs per nest (table 6.2). The multiple regression analysis shows that of the four independent variables, only the size difference between mates explains a significant amount of the variation in the number of unhatched eggs (table 6.2). This is true whether the effect of this variable is considered before or after the effect of the other variables is removed.

I also determined whether the size difference between mates explained a significant amount of the variation in the reproductive output of the female (as opposed to the loss of eggs just analyzed). The four independent variables were the same as in the above analysis, but here the dependent variable is

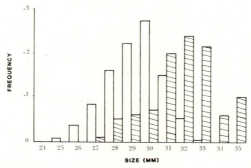

Figure 6.6 The frequency distribution of the sizes of male (open bars) and female (striped bars) *P. pustulosus* at Kodak Pond in 1979 (from Ryan 1983b).

Table 6.2 Various Factors Influencing Number of Unhatched Eggs

	df	ss	F	P
$R^2 = .22$				
Regression	4	39.99	4.54	< .005
Residual	63	138.63	—	—
SS Explained by Variable in the Order Listed				
Regression	4	39.99	—	—
MS	1	2.26	1.03	> .100
FS	1	2.25	1.02	> .100
CS	1	1.55	0.70	> .100
MSD	1	33.93	15.41	< .005

Notes: MS = male size; FS = female size; CS = clutch size; MSD = absolute value of the size difference between mates on the number of unhatched eggs.

the number of eggs hatched. Not surprisingly, clutch size explains most of the variation in the number of eggs hatched. Because female size and clutch size are correlated, only the variable that is considered first in this pair explains a significant portion of the total variance (table 6.3). In whatever order the variables are considered, the size difference between mates still explains a significant amount of the variation in the number of eggs per clutch.

Table 6.3 Various Factors Influencing Number of Hatched Eggs

	df	ss	F	P
$R^2 = .98$				
Regression	4	6414551.1	5829.3	< .005
Residual	63	6932.7	—	—
SS Explained by Variables in the Order Listed				
Regression	4	641451.1	—	—
CS	1	639489.2	5811.4	< .005
MS	1	151.6	1.4	> .100
FS	1	62.7	.6	> .100
MSD	1	1748.0	15.9	< .005

Note: See table 6.2.

It was necessary to determine whether unhatched eggs were a laboratory artifact; for instance, perhaps males did not fertilize eggs as efficiently in an unfamiliar environment. Since pairs often nest communally in the field, a nest might result from the reproductive effort of several pairs. I collected ten of these nests in the field the day after they were constructed. They were placed in the same laboratory buckets I had used in the earlier experiments. These nests contained a total of 3576 eggs, 5% of which did not hatch, a figure close to the 4% unhatched eggs in the laboratory experiments. This finding is also consistent with the frequent observations I made in the field of unhatched, and presumably unfertilized, eggs floating in the water near foam nests.

In some sense, that the fertilization process has not become perfected over evolutionary time is surprising. However, even in species with a relatively simple fertilization procedure, in which the male just hangs onto the female and releases sperm as she releases eggs, not all the eggs are fertilized (e.g., *Bufo bufo*, Davies and Halliday 1977; *Rana sylvatica*, Smith-Gill and Berven 1980; *Hyla rosenbergi*, Kluge 1981; *Bufo americanus*, Kruse and Mounce 1982). Therefore, we should not be surprised that fertilizations are missed in *P. pustulosus*, which exhibits a relatively complicated nesting procedure: the male uses his hind feet to guide the eggs from the female's cloaca, past his cloaca where they are fertilized, and then into the foam nest. Since the male can only handle several eggs at a time, he must repeat this process about a hundred times before the females extrudes her total clutch (see chap. 3). Not only are the mechanics relatively complicated, but this procedure is energetically demanding as well; males are operating at close to their maximum rate of oxygen consumption for the one hour or so during which nesting takes place (see chap. 7). Perhaps the suggestion of Licht, and Davies and Halliday that mates of similar size afford some mechanical advantage during fertilization is even more relevant for the complex nesting procedure of *P. pustulosus*.

Why Do Females Not Choose Mates of "Optimal" Size?

Most female *P. pustulosus* are larger than most males (fig. 6.6). Therefore, by mating with larger males, most females minimize the size difference between themselves and their mates. This mating preference may result in an increase in their reproductive output. Therefore, one could argue that natural selection favors female choice of larger males. However, some females are smaller than some males (fig. 6.6), and an optimality model would predict that smaller females should not select larger males, but should mate instead with males that are closer to their own size, thus minimizing the size difference between themselves and their mates. I measured the size of males and females of 65 nesting pairs collected at Kodak Pond to determine whether females mated assortatively by size, and especially to determine if smaller females avoided mating with the larger of the males. No significant correlation was found between the size of mated males and females ($r = -.10$, $P > .05$). I also divided males and females into small, medium, and large size classes (fig. 6.7). I noted how many matings took place between individuals of each size class, and found that females were assorted randomly with respect to the size distribution of the mated males

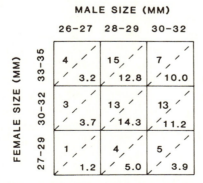

Figure 6.7 The observed (top of each cell) and the expected (bottom of each cell) number of matings between male and female *P. pustulosus* of each size class (from Ryan 1983b).

($X^2 = 2.56$, $df = 4$, $P > .75$). This also was true if small ($X^2 = 0.54$, $df = 1$, $P > .75$), medium ($X^2 = 0.54$, $df = 1$, $P > .75$), or large ($X^2 = 1.48$, $df = 1$, $P > .75$) size classes of females were considered separately. These results show that with respect to male size, females of different sizes do not exhibit different mating preferences—they all prefer larger males.

In terms of its effect on female fecundity, why smaller females choose larger males is not clear, since this choice should tend to reduce the females' reproductive output. However, several possibilities exist, two of which I will discuss. Borgia (1979) suggests that in some mating systems, selection may favor those females that evaluate both potential genetic and nongenetic benefits provided by males. If larger male *P. pustulosus* had better genetic quality (either due to an advantage through natural selection or sexual selection), most females would obtain both classes of benefits—increased reproductive output and better alleles for their offspring—by choosing larger males. This would not be true for smaller females, however; they would pass on better alleles to their offspring, but they would suffer an immediate loss in reproductive output. According to Borgia, small females should choose larger males if the genetic benefits afforded by these males compensate for the loss in reproductive output that results from mating with males that are larger than themselves. This model is difficult to evaluate in relation to *P. pustulosus* for the same reasons outlined in the previous discussion of eugenic mate choice—we just do not know how differential female mate choice influences the genotypes of their offspring. And again, the internal consistency of this theory needs to be evaluated more rigorously.

Another consideration involves neurophysiological constraints on female preference. For a female to choose mates based on a size relative to her own size, her preference must change with her size. We assume that the female preference for certain spectral characteristics of the chuck is somehow mediated by the response properties of her auditory system. Therefore, an ontogenetic change in call preference requires an ontogenetic change in the response properties of the auditory system. The response properties of the auditory nerve in bullfrogs (Fishkopf, Capranica, and Goldstein 1968) and toads (Capranica 1976a) show no changes that are associated with size. This does not necessarily mean that constraints exist on ontogenetic changes in the response properties of the auditory nerve in these (or in any) species. However, if we ask why a female mating preference that is based on an acoustic display does not change with female size, we are also asking questions about the ontogenetic flexibility of the underlying neural systems involved, and few data permit a consideration of this type of flexibility in the anuran auditory system.

Although the behavior that leads to preferential mating with larger males might decrease the reproductive output of some smaller females, it is interesting to consider quantitatively how this mating preference influences the reproductive output of females of all sizes. This can be determined from the probability that a female will mate with a male of a given size, which is a product of the size distribution of mated males and the probabilities of males of different sizes mating, and the expected number of eggs lost when a male and female of a given size mate (fig. 6.5). Therefore, the number of eggs lost due to the size difference between a female and her mate, for a female of size x is:

$$U_{fx} = (U_{|fx - mx|})(F_{mx}),$$

where $U_{|fx - mx|}$ is the number of eggs lost when a male and female of each size x mate, and F_{mx} is the frequency distribution of the size of mated males.

I solved this equation for females of each 1-millimeter size interval from 27 to 35 millimeters (fig. 6.8). The graph shows the expected number of eggs lost because a female is larger than her mate (i.e., the male is too small), and because a female is smaller than her mate (i.e., the male is too large). As expected, smaller females lose more eggs because their mates usually are too large. At 30 millimeters, most of the eggs are lost because males are too small, and virtually all of the egg loss experienced by females 31 millimeters and larger occurs because the females are larger than their mates. Summed for all female sizes, the model shows that a significantly greater loss in reproductive output is due to females being larger than their mates (mean 7.0 eggs lost), as opposed to females being smaller than their mates (mean 2.6 eggs lost; Mann-Whitney U test, $U = 19$, $P < .05$). Therefore, although females choose larger males as mates, and this choice might actually reduce the reproductive output of smaller females, most eggs are lost because even these larger males are too small to maximize the rate of fertilization of most females.

The above model actually overestimates the reproductive loss due to small females mating with large males, because it does not take into account the fact that relatively few small females are found in the population. This

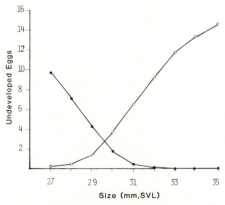

Figure 6.8 $U_{fx} = (U_{|fx - mx|})(F_{mx})$ solved for females of each size. The open circles show the expected number of eggs lost (U_{fx}) due to a female mating with males smaller than herself, and the closed circles U_{fx} due to a female mating with males larger than herself.

shortcoming can be remedied by entering the frequency distribution of female sizes (from fig. 6.6), F_{fx}, into the equation, yielding:

$$Ub_{fx} = (U_{|fx - mx|}) (F_{mx}) (F_{fx}),$$

where Ub_{fx} is a relative measure of eggs lost by a female of size x which is adjusted for the proportion of females of size x in the population.

The results of this equation show that this relative measure of reproductive loss for smaller females is not much greater from mating with males too large than it is from mating with males too small (fig. 6.9). When we compare this measure of loss of reproductive output due to mating with males that are larger versus mating with males that are smaller for the entire population, we see that relative reproductive loss is an order of magnitude greater when females are larger than their mates (1.0 versus 0.1; Mann-Whitney U test, $U = 15$, $P < .025$). Of course, these conclusions must be tempered by the observation that a large variance exists in the average number of eggs lost between mates of a given size difference (fig. 6.4).

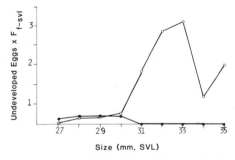

Figure 6.9 $Ub_{fx} = (U_{|fx - mx|}) (F_{mx}) (F_{fx})$ solved for females of each size. The open circles show the expected number of eggs lost balanced for the proportion of females of that size (Ub_{fx}) due to a female mating with males smaller than herself, and the closed circles show Ub_{fx} due to a female mating with males larger than herself.

On average, females should select larger males as mates to decrease egg loss, even if this is of some disadvantage to smaller females, because the frequency distribution of female sizes used in the model also represents the probability that an average female will breed at each size. The results of the model can be interpreted as a demonstration that over the life history of an average female, she will suffer greater loss of reproductive output because she is mating with males that are too small as opposed to mating with males that are too large. If a mating preference for male size must be fixed, then it makes sense in terms of natural selection for females to select larger males. However, this model contains two assumptions that suggest caution: (1) No data on the reproductive patterns (e.g., probability of breeding at each age) for individual females exist. Although we can estimate the reproductive pattern of an "average" female, we have no idea of the magnitude of the variance. (2) It is only an assumption that an individual female's mate preference is fixed and does not change with her size because of neurophysiological constraints.

One more point in relation to female life history patterns must be considered. In *P. pustulosus*, as in most anurans (Salthe and Duellman 1973), a positive correlation exists between female size and clutch size (fig. 6.3). Because female reproductive output, and presumably fitness, increases with size, there should be selection for large female size. Possibly, however, the male size distribution might impose a ceiling on the maximum female size: as females get larger, more eggs are lost, because the size difference between the female and her potential mates also increases. This is shown clearly in figure 6.8, where a female of 35 millimeters should, on the average, have a much greater loss of reproductive output from mating with males smaller than herself then would a 27 millimeter female. It seems unlikely, though, that the male size distribution will impose significant constraints on female body size, for the following reason: Egg loss from mating with smaller males never increases by more than 4 or 5 eggs for any 1 millimeter increase in female size (fig. 6.8), but for a similar increase in female size her clutch size increases by almost 100 (fig. 6.3). This increased clutch size will more than compensate for any lost reproductive output due to the relative decrease in the size of potential mates.

The data presented in this section indicate that an average female suffers a 4% loss in reproductive output, due in part to the smaller size of the male with which she mates. Although 4% may appear an insignificant fraction of a female's total reproductive output, under certain circumstances it could provide enough variation for selection to act upon. If there were a heritable component to the female mating preference, and if this mating preference influenced the female's reproductive output by as little as 1% relative to females without the preference, this preference should spread through the population. Fisher spoke of the possibility of selection acting on such relatively small advantages:

> If we speak of a relative advantage of one per cent, with the meaning that animals bearing one gene have an expectation of offspring only one per cent greater than those bearing the allelomorph, the selective advantage in question will be a very minute one; at least in the sense that it would require an enormous number of experimental animals, and extremely precise methods of experimentation, to demonstrate so small an effect experimentally. Such a selective advantage, however, would greatly modify the genetic constitution of the species, not in 100,000 generations but in 100 generations! (1958, p. 5)

The main concern in applying the nongenetic-benefit argument to the evolution of female choice is not the ability of selection to act on a 4% differential, but the large variance in the number of eggs fertilized as a function of the size difference between the pairs.

The Evolution of Female Choice in *Physalaemus pustulosus*

This chapter has shown that female mate choice can have a small influence on female fecundity. Thus possibly mate choice in this species has been influenced by natural selection. However, the inescapable conclusion drawn from chapter 5 is that female choice is responsible for the evolution of the chuck component of the advertisement call, and that it might also have been responsible for the evolution of the relatively low frequencies of the call. This conclusion strongly argues for the evolution of male traits under the influence of sexual selection. Natural selection through either a genetic effect (selection for increased growth and survivorship) or a nongenetic effect (selection on female fecundity) might have been the force that originally selected for females to choose larger males with lower frequency calls. However, the phylogenetic analyses suggest that sexual selection on the male call then came into play and probably resulted in the joint evolution of the male trait and the female preference in a Fisherian process.

As suggested by Fisher (1958), an evolutionary change should occur in those male traits that enhance mating success, in this case male size or the frequency of the chuck. As a result of the genetic correlation of the alleles that influence the female preference and the male trait, selection or stochastic factors affecting one will lead to a correlated genetic response in the other (Lande 1980, 1982; Kirkpatrick 1982). Because of the potentially rapid evolutionary response of male traits to female choice, over evolutionary time female preference might result in changes over relatively short time intervals after genetic variation arises. During other times, the female preference might be based on nonheritable variation in the male trait (West Eberhard 1979), or the female preference might be maintained, but not be exercised, in the absence of sufficient variation in the male trait (Anderrson 1982a).

The evolutionary scenario is quite different if females only receive a nongenetic benefit. The female preference can evolve in the absence of a heritable component of male fitness. Evolution of the female preference will proceed through the selective advantage of increased reproductive output achieved by discriminating females; however, no evolutionary response of the male's trait to female mate choice will occur. However, this seems unlikely for túngara frogs given the phylogenetic analysis presented in the last chapter that argues for an evolutionary response of the male trait to female choice.

7 Costs of Reproduction: Energy

Teleological approaches to the evolution of behavior assume that a behavior has evolved under the influence of selection acting to maximize the fitness of the individual. However, this is true only under several assumptions. The first is that the behavior being considered displays heritable genetic variation. Only rarely does information exist about the heritability of behavior (see Arnold 1981 for a notable exception); this limitation was made clear in the previous chapters. The second assumption is that no constraints exist (e.g., morphological or developmental) on the evolution of the behavior in question. An example of one such constraint from this study is the influence of vocal sac size on the lower frequencies in the advertisement call. A final assumption is that of "everything else being equal," which is probably never the case. In reproductive behavior, for example, both increased energy expenditures and exposure to predation are two "costs" often associated with sexual displays that might otherwise increase a male's ability to attract mates (see Lewontin 1977 for a detailed consideration of this subject).

I do not mean to suggest that optimization approaches to the study of adaptation are fundamentally flawed, only that constant attention must be given to the above assumptions (e.g., Oster and Wilson 1978; J. Brown 1982; Mayr 1983). Special emphasis has been placed on the costs of reproduction (e.g., see Williams 1966), and the analyses of these costs are a significant aspect of theories of sexual selection (e.g., Fisher 1958; Trivers 1972) and life history evolution (e.g., Williams 1966; Stearns 1976). This chapter and the next present quantitative estimates of two of the costs of reproduction incurred by túngara frogs: energy expended by both sexes in reproduction, and risk of predation experienced by males while advertising for mates. I will consider how these costs might have influenced the evolution of reproductive behavior in this species.

A basic assumption of most evolutionary studies of reproductive energetics is that animals have a limited amount of energy, only part of which can be devoted to reproduction, and the manner in which this energy is expended

143

is a product of natural selection. Most studies of the reproductive energetics of oviparous species have been limited to measures of the chemical potential energy of the eggs. These data often form the basis for theoretical considerations of the evolution of reproductive strategies in various taxa (e.g., Tinkle, Wilbur, and Tilley 1969; Tinkle and Hadley 1975). However, this analysis tends to ignore other constraints on reproductive patterns of females or the considerable amount of energy that females expend in other aspects of reproduction. For example, lizards of the genus *Anolis* are arboreal and have toe pads that aid in climbing. Andrews and Rand (1974) suggested that the structure of the toe pads limits the weight a female can carry and still climb effectively. They suggested that this limitation might be primarily responsible for the clutch size of only one egg in these lizards. Also, Rand and Rand (1976) showed that the amount of energy expended during nest digging is an important constraint on the reproductive biology of female green iguanas.

Due to the difficulty of measuring energy expended by males engaged in reproductive behaviors, it is almost never possible to compare the energy expended for reproduction by two sexes of the same species. This omission is important, since the information is crucial for interpreting patterns of sexual dimorphism or sexual selection in any species (e.g., Trivers 1972).

Energy Expended during Calling

The primary reproductive behavior of male túngara frogs is calling. Therefore, a measure of the energy that males expend during calling is a good approximation of the total amount of energy they devote to reproduction.

Together with George Bartholomew, Terry Bucher, and Stanley Rand, I measured the oxygen consumption, or aerobic metabolism, of the frogs while they were calling. Because males called for much of the night, we assumed that this behavior was supported primarily by aerobic, as opposed to anaerobic, metabolism. Usually activities that are supported anaerobically result in substantial buildup of lactic acid, and such activities cannot be maintained for long periods of time. Male *P. pustulosus* called in respirometers made from large peanut-butter jars. The methods of measuring the amount of oxygen consumption by males while they were calling are given in detail in the appendix. Briefly, we knew the fractional concentration of oxygen present in respirometers of the same size with and without a frog. By comparing the amount of oxygen in the two respirometers after a period of calling, we could determine how much oxygen the frog utilized during this period. Given the amount of oxygen consumed by the frog at rest, and the number of calls produced by the frog, we also could determine how much oxygen the frog utilized for calling in excess of the amount of oxygen utilized for maintenance. For comparative purposes, we also measured the rate of oxygen consumption at rest during the day, at rest at night, and when the frog received acoustical stimulation from the chorus (both the natural and the artificial chorus) but did not call.

The average rate of oxygen consumption for 22 frogs at rest during the daytime was 0.26 milliliters of oxygen per hour (table 7.1). The rate of oxygen consumption for frogs at rest during the night was more than twice that of frogs at rest during the day (table 7.1). Similar differences in rates of oxygen consumption as a function of time of day also have been reported for other anurans (Weathers and Snyder 1977; Carey 1979). A surprising result was that the rate of oxygen consumption of frogs that did not call, but received acoustical stimulation, was 1.3 times higher than the nocturnal resting rate. The rate of oxygen consumption during calling was 4.4 times the daytime resting rate, 2.1 times the nocturnal resting rate, and 1.6 times the stimulated rate.

We determined the incremental cost of calling for 9 frogs by subtracting their rate of oxygen consumption during nighttime resting from their rate of oxygen consumption during calling (table 7.2). This measure subsumes the increased oxygen consumption due to stimulation from the chorus. The incremental cost of calling varied greatly for these frogs, and we used a multiple regression analysis to determine how much of this variation was explained by the whine rate (i.e., the number of whines per hour), the chuck rate, and the frog's mass (table 7.3; fig. 7.1). The analysis shows that the whine rate explains 91% of the variance in oxygen consumption among

Table 7.1 Rates of Oxygen Consumption in *Physalaemus pustulosus*

	Mass (grams)	Oxygen Consumption (milliters/hour)			
	Mean (SE)	Mean (SE)	Max	Min	N
$\dot{V}O_2$ day	1.72 (.06)	0.26 (.01)	0.47	0.15	22
$\dot{V}O_2$ night	1.66 (.08)	0.53 (.08)	0.66	0.37	10
$\dot{V}O_2$ stim	1.76 (.10)	0.70 (.07)	0.96	0.28	10
$\dot{V}O_2$ call	1.67 (.06)	1.13 (.13)	1.94	0.62	9

Source: Bucher, Ryan, and Bartholomew 1982.
Note: $\dot{V}O_2$ day = oxygen consumption during daytime resting; $\dot{V}O_2$ night = nighttime resting; $\dot{V}O_2$ stim = acoustical stimulation but not calling; $\dot{V}O_2$ call = calling.

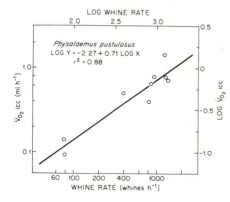

Figure 7.1 The incremental cost of calling ($\dot{V}O_2icc$; $\dot{V}O_2$ calling minus $\dot{V}O_2$ nighttime resting) as a function of rate of production of whines (equals the calling rate; from Bucher, Ryan, and Bartholomew 1982).

Table 7.2 Rates of Calling and Oxygen Consumption for Individual *Physalaemus pustulosus*

Mass (grams)	Whine Rate (per hour)	Chuck Rate (per hour)	$\dot{V}O_2$ Call (milliliters per hour)	$\dot{V}O_2$ Day (milliliters per hour)
1.57	844	0	1.18	0.23
1.51	797	267	0.92	0.21
1.64	1232	1108	1.94	0.27
1.57	78	8	0.62	0.24
2.05	76	20	0.67	0.47
1.81	922	786	1.31	0.26
1.60	396	140	1.02	0.20
1.77	1268	1904	1.30	0.30
1.48	1354	574	1.24	0.22

Source: $\dot{V}O_2$ day from Bucher, Ryan, and Bartholmew 1982.
Note: Rate of whine production equals calling rate.

Table 7.3 Factors Influencing Incremental Cost of Calling in *Physalaemus pustulosus*

	Proportion of R^2 Explained by each Variable
Whine rate (logarithm)	.8801
Body mass (logarithm)	.0788
Chuck rate (logarithm)	.0005

Note: R^2 is the total amount of the variation explained by all variables combined $R^2 = .91$.

males, and that the inclusion of the chuck rate does not explain significantly more of the variance than does the whine rate alone. Not surprisingly, the incremental cost of calling increases with the whine rate (fig. 7.1).

These results have several implications. One is that the rate of oxygen consumption during resting changes with the time of day, as is true for several other species of anurans. Thus studies that pool measurements of rate of oxygen consumption from different times of day (e.g., Seymour 1973; McClanahan, Stinner, and Shoemaker 1978; Hillman and Withers 1979) might obscure variability of biological importance. Interestingly, the rate of oxygen consumption of males increases when they hear a frog chorus, even if they do not exhibit an overt behavioral response to the acoustical stimulation. This means that males that do not call while at the breeding site, and even remain perfectly still, are still expending energy above that of resting due to their stimulated state. Noncalling males might be conserving energy relative to calling males, as suggested by Perrill, Gerhardt, and Daniels (1978, 1982), but the difference in energy expenditure between calling and noncalling males might be less than has been previously assumed.

The incremental cost of calling is the difference between the rates of oxygen consumption during nighttime resting and calling, and this increase in oxygen consumption results both from being stimulated and from calling. The amount of oxygen expended to produce a single call can be determined by dividing the incremental cost of calling by the whine rate. Surprisingly, the oxygen expended per call decreases with the calling rate (fig. 7.2). We are not sure why this is so, but we have suggested a simple hypothesis. If we assume that the oxygen expended due to being stimulated does not change with the calling rate, then the relationship shown in figure 7.2 is the result of changes in the mechanical cost of calling. When a frog calls, the air is shuttled back and forth between the lungs and the vocal sac, and the elastic nature of the vocal sac and the tissues of the lungs probably provide some of the mechanical force necessary for moving the air back and forth. We suggest that this shuttling of air probably expends less energy than a male's reinflation of his lungs by buccal pumping against tissue compliance. Thus the more calls that a male can produce with air obtained from a single inflation, the less the amount of energy expended for a single call.

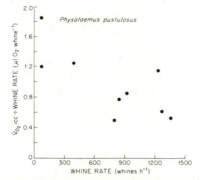

Figure 7.2 The incremental cost of calling ($\dot{V}O_2$ *icc*) for a single whine as a function of the rate of production of whines (equals the calling rate; from Bucher, Ryan, and Bartholomew 1982).

However, not all the energy expended during calling results from aerobic metabolism. Sometimes the amount of energy required for a behavior exceeds that which can be marshaled by oxygen consumption. In these situations, an oxygen debt is built up through the process of anaerobic metabolism and must later be paid back through aerobic metabolism. Lactic acid is the end product of this process, so the amount of lactic acid is indicative of the magnitude of the oxygen debt. Therefore, we determined the amount of lactic acid present in the bodies of males after prolonged calling. For comparative purposes, we did the same for resting and nesting individuals of both sexes.

Table 7.4 Lactic Acid Concentrations and Behavior in *Physalaemus pustulosus*

Behavior	Sex	N	Mass (grams)		Lactate (milligrams/gram)	
			Mean	SD	Mean	SD
Resting	M	4	1.70	.25	.32	.22
	F	4	1.71	.38	.34	.31
Nest building	M	4	1.49	.20	.65	.24
	F	4	1.60	.22	.44	.13
Continuous calling	M	5	2.18	.28	.50	.21
Sustained activity (35 minutes)	M	4	1.39	.10	.67	.25

Source: Ryan, Bartholomew, and Rand 1983.

The average amount of lactic acid in resting males was 0.323 milligrams of lactic acid for every gram of body weight, and 0.340 milligrams per gram for females (table 7.4). Five males called for an average of 3.6 hours, and the average lactic acid concentration in the bodies of these males was 0.503 milligrams per gram (table 7.4). The difference between the concentrations of lactic acid in resting males and calling males was not statistically significant (Student's t test; $t = 1.28$, $P > .20$). We assume that production of one milligram of lactic acid produces 0.017 millimoles of adenosine triphosphate (ATP, the molecule whose oxidation releases energy), and that consumption of one milliliter of oxygen produces 0.290 millimoles of ATP. Therefore, during 3 hours of calling, aerobic metabolism results in the production and oxidation of 0.988 millimoles of ATP, while anaerobic metabolism only yields 0.018 millimoles of ATP. The contribution of anaerobiosis to calling is minimal (table 7.5).

The Efficiency of Converting Metabolic Energy to Acoustic Energy

Males expend a considerable amount of energy during calling. Calling is necessary to achieve matings, and as will be discussed later, the amount of energy available to males might limit the number of nights they visit the breeding site. Therefore, the efficiency with which males convert metabolic energy to acoustic energy is an important aspect of the species' reproductive biology.

The efficiency with which metabolic energy and acoustic energy are coupled is determined by comparing the amount of energy above maintenance used to produce a call with the amount of energy contained in the call. The amount of oxygen consumed per call varies with calling rate. The average amount is 1.2 microliters of oxygen per call (fig. 7.2), which converts to 0.024 joules of energy input per call.

The total power in a call is a product of the power at a point source and the area over which the call is radiated. I assume that the frog is an omni-

Table 7.5 Contribution of Aerobic and Anaerobic Metabolism (in ATP Yield) in Support of Various Behaviors of *Physalaemus pustulosus*

Behavior	Mass (grams)	$\dot{V}O_2$ (ml)	Lactic Acid (milligrams)	ATP Yield (millimoles) Aerobic	Anaerobic	Aerobic ATP Yield as % Total
M + F building nest	3.0	2.32	1.64	.68	.03	96
M calling for 3 hours	1.5	3.04	0.76	.99	.01	99
M sustained activity for 35 minutes	1.5	0.87	1.01	.40	.02	93

Notes: Physalaemus pustulosus body mass normalized to 1.5 grams. $\dot{V}O_2$ is the amount of oxygen in milliliters consumed during each behavior (from Ryan, Bartholomew, and Rand 1983).

directional source, and that water reflects most of the frog's call. Therefore, the call is radiated over a hemisphere. The median peak amplitude of calls with chucks measured at a distance of 50 centimeters was 90 dB SPL (chap. 4). Knowing the peak amplitude of the call, the mean amplitude can be determined by averaging the waveform of the entire call, and amplitude can then be converted to power (dB SPL = W_1/W_0, where $W_0 = 10^{-12}$ watts). Sinces watts = joules/second, the energy content of the call is determined by dividing watts by call duration. The total energy content of the call is then calculated from the area of the hemisphere over which the call is radiated, in this case 2π 0.5 m. Conversion efficiency is simply: (energy output/energy input) 100%.

Table 7.6 Power, Energy, and Energetic Efficiency for *Physalaemus pustulosus* Calls

Call	Total Power (mW)	Call Duration (sec)	Total Energy	Energetic Efficiency
Whine + 1 chuck	0.36	0.34	0.12	0.5%
Whine + 2 chucks	0.37	0.40	0.15	0.6%
Whine + 3 chucks	0.46	0.54	0.25	1.0%
Whine + 4 chucks	0.44	0.69	0.30	1.2%

Source: Ryan 1984.

Since energy input does not change with the number of chucks, but energy output increases, calls with more chucks are produced more efficiently (table 7.6). Energy coupling was very inefficient for all calls, ranging from 0.5% to 1.2%. However, the production of acoustic energy by animals might be a generally inefficient process. In the only comparable studies, Brackenbury (1977) estimated the efficiency of crowing in a cock to be 1.6%, and Mac-Nally and Young (1981) found that cicadas produce sounds with an efficiency of 0.8%. I suggest that sound production might always tend to be inefficient because frogs, birds, and cicadas all produce sounds in radically different ways.

One of the reasons why the coupling of metabolic and acoustic energy is so inefficient in *P. pustulosus*, and probably in other animals as well, should be apparent from my discussion of the properties of biological radiators in chapter 5. The wavelengths of túngara frog calls are much longer than the structures that radiate them, thus the efficiency of radiation, and by extension energy output, is low. Morphology, then, sets a lower limit to which frequencies can evolve and still be produced and transmitted effectively. This factor often is not considered in studies that predict the evolution of low frequency acoustic signals because these frequencies attenuate less in the environment (e.g., Morton 1975; but see Ryan and Brenowitz, 1984). Some tradeoff must exist among the efficiencies of energy conversion, radiation, and transmission.

Energy Expended during Nest Building

P. pustulosus was a good subject for the study of reproductive energetics, because not only would males call in respirometers, but mated pairs would construct nests there as well. Although the proportion of time that a male invests in nest construction is small relative to the time spent calling, nest construction appears to be a vigorous and energy-demanding activity.

Construction of the nest takes about one hour. It was not possible to measure the amount of oxygen consumed by both individuals separately, since they both had to be in the same respirometer. In our calculations we divide the total amount of oxygen consumed equally between the male and the female. Since the male is engaged in more vigorous activity during nesting, we probably underestimated the male's rate of oxygen consumption and overestimated that of the female. The rate of oxygen consumption ranged from 1.76 to 3.54 milliliters per hour. The average rate for an individual was 1.32 milliliters per hour, which as 1.2 times the rate of oxygen consumption during calling, and 1.9, 2.5, and 5.1 times the rates during acoustic stimulation, nighttime resting, and daytime resting, respectively (table 7.7).

Table 7.7 Oxygen Consumption by Nesting Pairs of *Physalaemus pustulosus*

Mass (grams)		Duration	Rate of Oxygen Consumption
Male	Female	(minutes)	(milliliters/hour)
1.99	2.23	58.75	2.567
1.64	1.41	79.20	2.599
1.77	1.79	57.50	2.912
1.83	1.76	83.25	1.909
1.98	2.08	38.58	3.540
1.24	1.37	49.25	1.764
1.72	1.10	52.00	2.449
1.54	1.28	66.50	2.706

Note: Oxygen consumption is divided equally between the male and female.

We also measured the concentration of lactic acid in the bodies of males and females immediately after nesting. The average concentration was 0.65 milligrams per gram for males and 0.45 milligrams per gram for females (table 7.4). In each pair, the male exhibited a higher concentration of lactic acid than the female, which supports our impression that males expend more energy during nesting than do females. The difference in lactic acid concentrations of nesting males and females were nearly statistically significant ($t = 2.04$, $.05 < P < .10$), but the concentrations in nesting males were not very different than those of calling males ($t = 0.99$, $P > .20$; table 7.4). As with calling, the contribution of anaerobic metabolism to ATP production during nesting was minimal (table 7.5).

Physiological Limits of Energy Expenditure

Only a certain amount and intensity of activity can be sustained by aerobic metabolism. Therefore, the capacity for sustained activity could be an important physiological constraint on the animal's reproductive behavior. Interestingly, this capacity differs considerably among species of frogs. Some researchers have suggested that the aerobic metabolic scope, which is defined as the maximum rate of oxygen consumption divided by the rate of oxygen consumption during resting, is correlated with certain behavioral features of the species (Bennett and Licht 1974). We determined the rate of oxygen consumption and the lactic acid concentration in frogs after periods of continuous, forced activity, thus allowing us to compute the aerobic metabolic scope for *P. pustulosus*. This measure serves two purposes: (1) it facilitates comparison of our data with data from other studies, since these studies usually do not measure voluntary, but only forced, activity; and (2) it provides some estimate of the physiological limits of a continuous reproductive behavior, such as calling, in túngara frogs.

The rate of oxygen consumption was determined under two conditions of forced activity. During the first, a frog was placed in a cylinder that was rotated through 180 degrees in 0.2 seconds and then held motionless for 1.8 seconds. The frog clung to the wall, fell to the bottom, and righted itself as the cylinder rotated. This was continued until the frog no longer righted itself, the average of which was 34.9 minutes. We also subjected frogs to a regime that required continuous activity: the cylinder was rotated continuously, leading to shorter periods of rest and an earlier cessation of righting attempts. Under this regime the frogs continued to right themselves for an average of 2.9 minutes (table 7.8).

The rate of oxygen consumption was 1.75 milliliters per hour during the 35 minutes of forced activity, and 3.35 milliliters per hour during the 3 minutes of continuous activity (table 7.8). The lactate concentrations of animals subjected to 35 minutes of forced activity was 0.67 milligrams per gram. This was significantly higher than the lactate concentration in resting

Table 7.8 Average Rate of Oxygen Consumption for Various Activities

Activity	$\dot{V}O_2$ (milligrams/ gram-hour)	Metabolic Scope (milligrams/ gram-hour)	Factorial Scope	Percentage at Maximum $\dot{V}O_2$
Daytime resting	0.15	—	—	8.2
Nighttime resting	0.32	0.17	2.13	17.6
Stimulated	0.40	0.25	2.67	22.0
Calling	0.68	0.53	4.53	37.3
Nesting	0.77	0.62	5.13	42.3
35-minute forced activity	1.16	1.01	7.73	63.7
3-minute continuous activity	1.82	1.67	12.13	—

Notes: $\dot{V}O_2$ = oxygen consumption; $\dot{V}O_2$ activity − $\dot{V}O_2$ daytime resting = metabolic scope; $\dot{V}O_2$ activity ÷ $\dot{V}O_2$ resting = factorial scope.

frogs ($t = 2.13$, $P < .02$), but it did not differ from the lactate concentration in calling frogs ($t = 1.5$, $P > .10$), or nesting frogs ($t = 0.99$, $P > .40$). During the 3 minutes of continuous, forced activity the rate of oxygen consumption was higher than in the longer (35 minutes) period of forced activity.

Comparison of Aerobic Capacity of Túngara Frogs with That of Other Anurans

A measure that helps to relate the rate of oxygen consumption to the behavior and ecology of the animal is its metabolic scope. This is a measure of the maximum rate of aerobic energy metabolism, or rate of oxygen consumption, above that needed for physiological maintenance, that can be marshaled by an animal in support of activity. It is calculated as the difference between the observed maximum and minimum rates of oxygen consumption. The metabolic scope, to a large degree, sets the physiological limit for the types and durations of behaviors used by animals. The factorial scope is the maximum rate of oxygen consumption divided by the minimum rate. Factorial scope is commonly used for interspecific comparisons, but this measure can also be used to compare the rates of oxygen consumption during different behaviors of túngara frogs. Here, I will speak of the factorial scope for different behaviors, and I will define the ratio of the maximum rate of oxygen consumption to the rate of oxygen consumption during resting as the maximum factorial scope.

The rate of oxygen consumption increases substantially during the reproductive behaviors of túngara frogs (table 7.8). The factorial scope ranges from 2.67 for noncalling males receiving acoustical stimulation from the chorus, to 5.13 for an individual of a nesting pair. The maximum factorial scope is 12.13, and occurs during short (3 minutes) bouts of forced activity.

We can see that during the most physiologically demanding reproductive behavior—nesting—frogs are operating at 42% of their maximum aerobic capacity.

Although many measures of metabolic scope for anurans have been made, only one study has measured the rate of oxygen consumption during voluntarily sustained activity. Seymour (1973) measured rates of oxygen consumption in the spadefoot toad (*Scaphiopus hammondi*) during resting, forced activity, and voluntary burrowing. The rates of oxygen consumption (milliliters per gram-hour) were 0.09, 0.85, and 0.36, respectively (these measures were adjusted to 25° C, assuming that metabolic rate doubles for every 10° C increase in body temperature). Thus *S. hammondi* has a maximum metabolic scope of 0.76 milliliters per gram-hour and a maximum factorial scope of 9.5. The rate of oxygen consumption during burrowing was 0.27 milliliters per gram-hour greater than resting, or 4.0 times resting, which is 35% of the toad's maximum rate of aerobic metabolism.

Some similarities exist between the results of the energetic studies of *P. pustulosus* and *S. hammondi*. Rates of oxygen consumption were measured during activities that the animals normally sustain for prolonged periods of time (burrowing, calling, and nesting), as opposed to rapid bursts of forced activity. The factorial scopes for calling and nesting by *P. pustulosus* are higher than the scope for burrowing by *S. hammondi*, but during these activities both species operate at almost the same relative level (42% versus 35%) of their maximum aerobic capacity. Another similarity is that both species have relatively high maximum factorial scopes for anurans. Bucher, Ryan, and Bartholomew (1982) reviewed previous measures of maximum factorial scopes from ten studies of anurans (all measurements were adjusted to 25° C). The scopes ranged from 3.9 to 22.2. Túngara frogs had higher factorial scopes than 7 of the 9 other species, and spadefoot toads had scopes higher than 5 of the other 9 species. Taigen, Emerson, and Pough (1982) measured oxygen consumption during resting and forced activity for 17 species of frogs at 20° C. Adjusting for the difference in temperature, *P. pustulosus* would have a factorial scope of 11.7 at 20° C, which is higher than 9 of the 17 species measured by Taigen, Emerson, and Pough. Not only does *P. pustulosus* have a relatively high maximum factorial scope, this species also has the highest maximum rate of oxygen consumption per gram (i.e., rate of oxygen consumption during forced activity) of any of the 23 species reviewed by Bucher, Ryan, and Bartholomew (1982) and by Taigen, Emerson, and Pough (1982). In fact, the rate of oxygen consumption per gram during the voluntary activities of calling and nesting by túngara frogs is higher than the rate of oxygen consumption during forced activities for most frogs measured.

Since metabolic capacity has an important influence on an animal's behavior, it would not be surprising if these two factors were correlated. Bennett and Licht (1974) presented data suggesting such a correlation between physiology and behavior. They measured lactate production during forced activity in several species of amphibians, and found that the manner in which animals avoid predators was correlated with the degree to which the animals depended on aerobic versus anaerobic metabolism. Animals that used rapid bursts of activity to flee predators had a high anaerobic dependence, while those that relied on noxious secretions to avoid predators had a relatively low dependence on anaerobic metabolism. Bennett and Licht also suggested that other activities, such as the type of courtship, might be correlated with metabolic activity.

Taigen, Emerson, and Pough (1982) measured maximum rates of oxygen consumption and lactate concentration during forced activity in 17 species of frogs. They devised two measures of metabolic activity. The first—metabolic performance—indicates the total power input derived from aerobic and anaerobic metabolism during forced activity relative to other species. Larger values of metabolic performance indicate greater total power input. The second—aerobic dependence index—is a relative measure of the extent of dependence on aerobic metabolism, as opposed to anaerobic metabolism, during forced activity. Larger values indicate greater aerobic dependence. Taigen, Emerson, and Pough suggested that aerobic dependence was not a phylogenetically conservative character, and that it did not correspond to their strictly dichotomous classifications of habitat type or the manner in which the animals avoid predators. However, they did find that aerobic dependence was correlated with mode of locomotion and foraging behavior. They also suggested that other factors, such as reproductive behavior, might be correlated with this measure of the animal's dependency on aerobic metabolism.

The measures of aerobic dependence by Taigen, Emerson, and Pough (1982) ranged from -2.10 to $+1.96$. For comparative purposes, I adjusted the average maximum rate of oxygen consumption and lactate production for *P. pustulosus* to 20° C, the temperature at which Taigen, Emerson, and Pough conducted their experiments. It should be cautioned that because oxygen consumption was measured during 3 minutes of forced activity, and lactate production during 35 minutes of forced activity, the data from túngara frogs are not strictly comparable to those of Taigen, Emerson, and Pough. The data from my experiments probably overestimate lactate production, and this in turn would actually underestimate the dependence of túngara frogs on aerobic metabolism. The aerobic dependency for túngara frogs was $+3.60$, or 1.8 times the highest measure obtained by Taigen, Emerson, and Pough.

Taigen, Emerson, and Pough (1982) found that frogs that jump tend to have lower measures of aerobic dependence than ones than hop (jump and hop are defined by Emerson 1978). Túngara frogs hop, so their high measure of aerobic dependence is consistent with the results of Taigen, Emerson, and Pough. They also found that active foragers—frogs that move about in search of prey—had higher measures of aerobic dependence than passive foragers, which sit and wait for prey to pass by. Although they did not have data on foraging behavior for most of these species, they classified species from literature data on stomach contents. They assumed that animals with small prey from one or a few taxa are active foragers, while frogs that had taken a variety of larger prey items are passive foragers.

How do túngara frogs fit this prediction? I collected 29 male frogs from a chorus in Gamboa, and I analyzed the contents of their stomachs. Thirteen of these males had empty stomachs. The others had a variety of prey items, but they seemed to feed mostly on ants and termites (table 7.9). On the nearby Pipeline Road, ants and mites comprised 32% of the potential prey items in the leaf litter (Toft 1981). Túngara frogs eat a disproportionate amount of these prey items (60%), which suggests that they forage actively to locate their prey. Toft does not present data for abundance of termites in the leaf litter, but the túngara frogs probably could only eat such a high proportion of termites if they actively searched for them. Therefore, on the continuum from active forager to passive forager, *P. pustulosus* appears to be closer to the former. My data on aerobic dependency together with the estimate of foraging behavior are consistent with the suggestion by Taigen, Emerson, and Pough that active foragers have a higher dependence on aerobic metabolism.

Table 7.9 Proportion of Prey Items in *Physalaemus pustulosus* Stomachs

| Mites | Ants | Termites | Diptera | | Snail | Other |
			Adult	Larvae		
.33	.11	.11	0	.11	0	.11
0	1.0	0	0	0	0	0
0	.86	0	.08	0	0	0
0	0	0	0	0	1.0	0
0	0	1.0	0	0	0	0
0	0	1.0	0	0	0	0
0	.89	0	0	0	0	.11
0	1.0	0	0	0	0	0
0	.25	.75	0	0	0	0
0	1.0	0	0	0	0	0
0	.33	0	0	0	0	.66
.20	.60	0	0	0	0	0
0	1.0	0	0	0	0	0
0	1.0	0	0	0	0	0
0	0	1.0	0	0	0	0
0	1.0	0	0	0	0	0
Means						
.03	.57	.24	.01	.01	.06	.06

Note: Thirteen other animals sampled had empty stomachs.

Foraging is not the only behavior that might be correlated with aspects of the animal's physiology. Reproductive behavior, especially, should also be considered. I know of few species of frogs in which an individual male will call at such a rapid repetition rate, so continuously, and for so much of the night, as do túngara frogs. Also, few species of frogs make such a substantial expenditure of energy during oviposition as do túngara frogs during nest construction. Clearly the physiological capability of a male to sustain calling and nest building for prolonged periods of time should have a direct influence on his fitness. Thus it is not surprising that túngara frogs have a very high capacity for aerobic support of energy metabolism; it would be interesting to have similar data for other species with different reproductive behaviors. These observations also suggest that correlates of a frog's metabolic capacity cannot be explained simply by examining a small subset of the animal's behavior. If metabolic capacities are influenced by selection, then selection can be generated by a number of factors concerned with mating, feeding, avoiding predators, or a host of other aspects of the animal's behavior, ecology, and physiology. And if metabolic capacities prove to be evolutionarily conservative, then foraging is only one of many behaviors that metabolic capacity might constrain.

Energy Content of the Eggs

So far, I have considered the energy expended by males and females during active reproductive behavior. However, females invest a large amount of energy in their eggs. In order to compare energy expended for reproduction between the sexes, and also to compare túngara frogs to other species, we determined the mean chemical potential energy of five nests, four of which were deposited during laboratory experiments and one collected in the field. The nests were immediately dried in an oven at about 50° C for at least 24 hours. The energy content of the nests was determined with a bomb calorimeter at Cornell University in 1981.

The mean dry weight of the nests was 0.17 grams. In the field, the constituents of the nest are water, eggs, and sperm. The foam results from the uptake of water by the gelatinous coating of the eggs; the mass and the energy content of the sperm are negligible. Therefore, the nest is almost exclusively contributed by the female. The mean energy content of the nests was 3.96 kilojoules, or 22.91 kilojoules per gram of eggs. The energy content of the nest collected in the field (3.76 kilojoules) was within range of the values of the nests deposited in the laboratory.

Energy Expended by Male and Female Túngara Frogs during One Breeding Season

Using laboratory measures of energy expended for reproduction by both sexes, the total amount of energy expended during an entire breeding season can be extrapolated by estimating such variables as the amount of time that males engaged in calling, the rate of mating by males, and the number of clutches produced by a female during one breeding season. At least some túngara frogs breed during every month of the year on Barro Colorado Island, but breeding is concentrated during the rainy season, approximately from April to mid-December. Therefore, we estimate that the length of the breeding season is 259 days (1 April to 15 December). Túngara frogs experience a large amount of predation while they are at the breeding site, and many males probably do not live through an entire breeding season. However, for comparative purposes we estimate the energy expended for reproduction by males during an entire breeding season.

The number of nights on which males visit the breeding site is quite variable. During the 152 consecutive days that I monitored the breeding site in 1979, males were present for an average of 7.2 nights over the average time span of 43 days during which they were known to be alive (see chap. 4). Therefore, the average male visited the breeding site on 16.7% of the nights he was alive. As pointed out in chapter 4, I could not be certain that males were not present at other breeding sites when they were not at Kodak Pond, but a survey of nearby sites indicated strong fidelity to Kodak Pond.

Males do not appear to have "noncalling strategies." Although some males called more than others on any given night, a high correlation was found between the amount of calling by a male and the amount of time he spent at the breeding site ($r = .91$, $N = 617$, $P < .001$; chap. 5). Also, once a male began calling, most other males quickly joined in, and choruses stopped almost synchronously. These data suggest that no large differences exist in the amount of calling among males at the breeding site. Most of the calling at the breeding site occurred from 1900 to 2400 hours (fig. 3.5); and when we quantified the number of calls produced by males, we found that the modal number of hours during which males called each night was 6 (table 3.1). We use the incremental cost of calling—the rate of oxygen consumption during calling minus the rate of oxygen consumption during nocturnal resting, 0.60 milliliters per hour—to estimate the amount of energy expended during calling. Although there were no totally silent males in the population, for comparative purposes we use the increased rate of oxygen consumption that results from acoustical stimulation alone—the rate of oxygen consumption while stimulated minus the rate of oxygen consumption during nocturnal resting, 0.17 milliliters per hour—to estimate the amount of energy that would be expended by silent males.

The average male mated on 19% of the nights he spent at the breeding site. We use this percentage and the average number of nights a male is present at the breeding site to estimate the average frequency of nesting. The average rate of oxygen consumption for each individual of a nesting pair was 0.80 milliliters per hour above the nocturnal resting rate. This value probably underestimates the energy expended by males during nesting, and it overestimates that of females. However, we use this value to estimate energy expenditure by both sexes during the approximately one hour of nesting.

Female túngara frogs produce more than one clutch of eggs per season (chap. 3), but we lack sufficient data from this study to estimate the interclutch interval of females in the field. However, Davidson and Hough (1969; see also chap. 3) found that female túngara frogs reproduced every four to five weeks in the laboratory. The limited field data that are available are more consistent with an estimate of four weeks. We estimate that a female reproduces ten times in a breeding season, and we use the average chemical potential energy of a clutch of eggs to estimate the amount of energy expended for producing eggs over an entire season.

In order to compare amount of energy expended for egg production to that of reproductive behaviors, we converted oxygen consumption to joules by assuming that the consumption of 1 milliliter of oxygen yields 20.10 joules. Because anaerobic metabolism accounts for a negligible amount of the energy produced during reproductive behaviors, its contribution was ignored.

If a male lived through an entire breeding season, he would visit the breeding site on 44 nights, calling for an average of 264 hours. He would use an average of 3.12 kilojoules for calling. He would mate with females on an average of 8.4 nights, which results in an energy expenditure of 0.13 kilojoules. Over an entire season, an average male would expend 3.25 kilojoules for reproduction (table 7.10). If noncalling males were in the population, and if they mated at the same rate as calling males, they would expend 1.02 kilojoules for reproduction (table 7.9).

The energy expended by a female for the production of one clutch of eggs surpasses the amount of energy that an average male expends for reproduction in an entire year (table 7.10). If a female does produce 10 clutches of eggs per season, as we estimate, she would expend 41.0 kilojoules for reproduction (egg formation and nesting) during an entire breeding season.

Table 7.10 Estimated Seasonal Aerobic Energy Expenditures for Reproduction by Male and Female *Physalaemus pustulosus*

	Males		*Females*
Days present at breeding site	44		Energy per clutch 3.96
Hours present at breeding site	264		Energy for nesting 0.13
Energy for calling males	3.12		
Energy for silent males	0.88		
Nights mated	8.4		
Energy for nest	0.13		
Total energy expenditures			Total energy expenditures
Silent males	1.02		All females 41.0
Calling males	3.25		

Source: Ryan, Bartholomew, and Rand 1983.
Notes: Mean male mass = 1.7 grams; mean female mass = 1.8 grams. All energy expenditures are in kilojoules.

Thus an order of magnitude more energy is expended by females for egg production alone. Even if a male visited the breeding site and called all night on every night of the breeding season, and mated at the average rate, he still would expend only 24.90 kilojoules per season for reproduction, considerably less than that expended by a female. These data, which are the first for any frog species, allow a comparison of the energy expended for reproduction between the sexes and demonstrate that females invest much more energy in reproduction than do males; seemingly no amount of increase in reproductive behavior by the males could offset this difference.

Although males expend a small amount of energy for reproduction relative to females, some important constraints on male reproductive behavior which are associated with energy expenditure may still exist. That males are present at the breeding site for only a small proportion of the nights they are known to be alive and that males probably do not feed while they are at the breeding site (table 7.9) suggest that males cease reproductive behaviors in order to replenish energy supplies. For several frog species, as I will discuss shortly, energy expenditures are known to exceed energy intake during periods of breeding.

Comparisons with the Reproductive Energetics of Other Amphibians and Reptiles

Few data available allow us to make any sort of comparison of the reproductive energetics of *P. pustulosus* with those of other species. Only Crump and Kaplan (1979) have measured the caloric content of frog eggs. Their mass-specific values for the energy content of eggs of 9 species of Neotropical hylid frogs were slightly higher than the values we obtained for the túngara frog (24.71 to 25.73 kilojoules per gram for hylid frogs versus 22.92 kilojoules per gram for *P. pustulosus*). How many times per season these female hylids reproduce is not known.

No other direct measures exist of energy expended by males during reproductive behavior. Changes in body weight and depletion of energy stores of males during the breeding season suggest that when a male is reproductively active, his energy expenditures are probably greater than his energy intake (e.g., Smith 1976). MacNally (1981) measured depletion of energy reserves in somatic tissue, fat bodies, and liver of two Australian frogs—*Ranidella signifera* and *R. parinsignifera*—during the breeding season. He estimated that males of these species expend 2.22 and 2.30 kilojoules for reproduction during breeding seasons of 60 and 120 days, respectively (table 7.11). Over an entire breeding season, male *P. pustulosus* expended more energy for reproduction than either of the species of *Ranidella*. But the *Ranidella* are smaller (0.72 grams versus 1.70 grams; see footnote b, table 7.11), and have a much shorter breeding season than *P. pustulosus* (table 7.11). If we adjust energy expenditure for differences in body size (by dividing kilojoules by grams$^{.75}$), the *Ranidella* males expended about 1.3 times more energy per gram-season, and 3 to 6 times more energy per gram-day than the male *P. pustulosus*. Unfortunately, MacNally (1981) did not measure the caloric content of the eggs.

Table 7.11 Total Energy Expended for Reproduction by Three Species of Ectothermic Vertebrates

		Kilojoules per			Length of	
Species	Sex	Season	Gram-season[a]	Gram-day[a]	Breeding Season	Mass Grams
Physalaemus pustulosus	M	3.25	2.19	.008	259	1.7
Physalaemus pustulosus	F	40.96	26.43	.102	259	1.8
Ranidella signifera[b]	M	2.20	2.88	.047	60	0.7
Ranidella parinsignifera[b]	M	2.30	2.99	.024	120	0.7
Uta stansburiana	M	21.50	8.78	.075	117	3.3
Uta stansburiana	F−1 yr	44.80	23.96	.204	117	2.3
Uta stansburiana	F>1 yr	45.40	21.52	.186	117	2.7

Source: Ryan, Bartholomew, and Rand 1983.
[a] Mass-specific energy expenditures were determined by dividing mass by grams$^{.75}$.
[b] Masses of *Ranidella* are not given in MacNally (1981), but the average snout-to-vent length for both species was 21.5 millimeters. We calculated mass using the equation $M + 6.0 \times 10^{-2} SV^{3.24}$, where SV is snout-to-vent length in centimeters (from Pough 1980).

A number of studies reported the caloric content of lizard eggs; the values presented by Tinkle and Hadley (1975) for 10 species of lizards range from 24.6 to 30.2 kilojoules per gram dry mass of eggs. However, Nagy (1983) actually presented estimates of energy expenditure for reproduction by both sexes of the lizard *Uta stansburiana*. Nagy was able to make these estimates by measuring the metabolic rate of lizards in the field using doubly labeled water. He found that females expended about 45 kilojoules for reproductive activities, depending on their age (table 7.11). This is similar to expenditures by female *P. pustulosus*, and these values are still similar after they are adjusted for differences in body mass. The breeding season of *P. pustulosus* is longer than that of *U. stansburiana*; consequently, the energy expenditure per gram-day is greater for the lizards when adjusted to the length of its breeding season (table 7.11).

A large difference was found in the amount of energy expended by males of *P. pustulosus* and *U. stansburiana*. The lizards expended in excess of 6 times the total energy that the frogs expended, and 4 times the energy on a per gram-day basis. The female *U. stansburiana* expended 3 times more energy than a conspecific male on a per gram basis, while the female *P. pustulosus* expended 12 times more energy than the male *P. pustulosus* on a per gram basis.

Probably both methodological and biological reasons exist for the sexual differences in energy expenditure found between *P. pustulosus* and *U. stansburiana*. Nagy's measurements for the lizards include all of the behaviors associated with reproduction, such as territorial defense. In *P. pustulosus* we measured energy expenditure during what are clearly the two dominant reproductive behaviors—calling and nesting. But other energy expenditures, such as migration to the breeding site, are involved with reproduction. However, biological differences between the males of each species probably explain most of the difference in the sexually dimorphic patterns of energy expenditure. Male *U. stansburiana* set up territories before the breeding season, and for several months these territories are defended continuously from conspecifics. Thus males not only engage in sexual display, they also patrol their territories and engage in aggressive encounters with other males. In contrast, male *P. pustulosus* do not reside in breeding territories, and in fact, they do not defend territories at all. Male túngara frogs maintain interindividual distances, but only when they are at the breeding site where, in contrast to the lizards, they are not present continuously. Thus the reproductive behaviors of male *P. pustulosus* are not as energetically demanding as those of male *U. stansburiana*, and nowhere near as demanding as those of female *P. pustulosus*.

Summary of Reproductive Energetics

The results presented in this chapter reveal a significant energetic cost associated with reproduction for both sexes of túngara frogs. The rate of energy expenditure by calling males is 4.5 times the rate when they are at rest during the day. Even if males visit the breeding site but do not call, their energy expenditure increases due to the stimulation of the chorus. The males that are able to attract females expend additional energy during nest construction, at a rate that is 12 times the daytime resting rate. Males only spend a small proportion of their nights at the breeding site, and they probably do not feed there. The need to acquire energy to support reproductive behaviors may limit the amount of time a male can devote to sexual displays.

The coupling of metabolic energy to acoustic energy is inefficient, ranging from 0.5% to 1.2% depending on the number of chucks. Efficiency is low, in part, because the frogs produce calls with wavelengths that are too long to be radiated efficiently. This might be true for many animals and points out the necessity of considering morphology as a constraint on the evolution acoustic communication.

Compared to other frogs, *P. pustulosus* exhibits a high rate of dependence on aerobic metabolism to supply energy for its behaviors. Although this high aerobic dependence is consistent with predictions that metabolic capacity should correlate with certain modes of locomotion and foraging, it also allows males to engage in high-energy-demanding reproductive behaviors.

Chapter Seven

The most interesting result from these estimates of reproductive energetics is the disparity in energy investment between the sexes. Although this disparity is especially true for túngara frogs, it also is true for one species of lizard and probably is the rule rather than the exception among vertebrates. In the following chapter, this measure of "cost" of reproduction will be combined with another "cost" important to túngara frogs—predation. I will then examine how these costs might influence patterns of male reproductive behavior and sexual dimorphism in this species.

8 Costs of Reproduction: Predation

When an animal communicates, whether by visual displays, vocalizations, or chemicals, the signaling animal becomes more conspicuous. In fact, in many circumstances selection will tend to increase the conspicuousness of an animal's signal. It is easy to see that an animal should have an advantage in making itself more conspicuous when it is attempting to attract a male or defend a territory. However, the increased conspicuousness that results from communication may sometimes be a disadvantage, and one such situation occurs when an animal's communication behavior increases its conspicuousness to predators. In general, the advantages and disadvantages of increased conspicuousness should be opposing selective forces on communication signals. Again, Darwin's doubt that many of these behaviors could evolve by natural selection comes to mind.

The role of predation has been given major consideration in theories dealing with the evolution of animal communication systems (e.g., Moynihan 1970; Smith 1977). The importance of predation in the evolution of mimicry complexes is generally accepted (e.g., Cott 1940), and Marler (1955; chap. 5) presented evidence for the convergence of structure and function in bird songs; such convergence has resulted, in part, from the force of predation. Predation has also been considered an especially important influence on the evolution of sexual displays. For instance, in his theory of runaway sexual selection, Fisher (1958; chap. 1) suggested that female choice leads to extreme development of male traits until this development is opposed by natural selection, and that predation seems to be the most likely counterselective force. There are data to support Fisher's suggestion. Several studies have demonstrated that in polygynous species, males suffer a higher rate of mortality than females, presumably because they engage in reproductive behaviors that increase their risk of predation (Selander 1965; Trivers 1976; Howard 1981; Schoener and Schoener 1982). Haas (1976) and Endler (1980) presented experimental evidence suggesting that a particular male trait that seems to be under sexual selection, in these cases the color patterns of fish, increases the predation risk to males. The most dramatic example of a predation cost to male sexual display behavior is provided by Cade (1975). He showed that calling male crickets attract acoustically orienting parasites that deposit their larvae on calling males, ultimately decreasing the males' reproductive success.

It is obvious to anyone who has heard a frog chorus that frogs are very conspicuous when they are calling. Prior to my studies with Tuttle, no data existed to document an increased predation risk for calling male frogs. Emlen (1976) and Howard (1981) present observations suggesting that territorial male bullfrogs are more likely to be eaten by snapping turtles, but whether the calling per se influences predation risk is not clear. Especially interesting to me was the observation by Jaeger (1976) that the large marine toad, *Bufo marinus*, seemed to be attracted to the calls produced by a túngara frog chorus—and he made these observations at Kodak Pond on Barro Colorado Island.

Bat Predation and Sexual Advertisement by Túngara Frogs

The initial discovery that the bat *Trachops cirrhosus* uses acoustic cues to hunt frogs had immediate implications for studies of the behavior, ecology, and sensory physiology of both the bats and the frogs. Tuttle and I have been investigating bat-frog interactions in great detail. Here I will consider only the small part of our research that investigates how bat predation influences the reproductive behavior of *P. pustulosus*.

The existence of an acoustically foraging predator suggests that male frogs, by increasing their ability to attract females, might also increase their risk of predation. Several behaviors that male frogs, in general, might engage in to increase their probability of mating include: (1) calling more intensely (Fellers 1979a); (2) calling more frequently (Whitney and Krebs 1975); (3) using call types that are more attractive to females (chap. 5); (4) calling continuously throughout the night; and (5) remaining at the calling site during disturbances. Call intensity and call repetition rate probably do not influence differential mating success in túngara frogs, because males usually call at the maximum possible intensity and call repetition rate. Therefore, these behaviors display very little variation on which female choice could act. Also, no evidence has been found that some of these behaviors listed (i.e., 4 and 5) influence male mating success at all, but this seems likely. Reproductive behavior among males is varied; and males exhibiting a certain subset of behaviors, such as producing calls with chucks, are more likely to attract mates. Therefore, if this subset of behaviors also increases the predation risk of males, then bat predation should be a selective force opposing the action of sexual selection. This is the theoretical basis for the following experiments.

We determined if male túngara frogs that called more frequently or in-

Costs of Reproduction: Predation 165

tensely were more likely to attract *T. cirrhosus*. The preference of the bat was tested in a paired-choice test in an outdoor flight cage (fig. 5.10). These bats were given a choice between a *P. pustulosus* whine call played at a repetition rate that was about normal and one that was slower than normal. All bats were more likely to be attracted to the call broadcast at the normal call repetition rate, and the overall preference for the normal call repetition rate was statistically significant (table 8.1). Similarly, we presented two bats with whine calls of different intensities. Both bats were attracted preferentially to the more intense call, and the overall preference was statistically significant (table 8.1).

Table 8.1 *Trachops cirrhosus* Response to *Physalaemus pustulosus* Calls

	Response		
	Slow Call Rate	Normal Call Rate	P
Bat #1	2	6	.144
Bat #2	0	8	.004
Bat #3	0	8	.004
		$X^2 = 26.1, P < .005$	
	Response		
	Less Intense Calls	More Intense Calls	P
Bat #1	0	7	.008
Bat #2	2	6	.144
		$X^2 = 13.1, P < .005$	

Notes: Slow call repetition rate = 1 call/3.2 seconds; normal call repetition rate = 1 call/1.6 seconds. Less intense calls = 74 dB SPL; more intense calls = 78 dB SPL.

After male *P. pustulosus* have been calling for a period of time, they tend to call at about the same, and the maximum, call repetition rate and intensity. However, these experiments with bats show that if the frogs did call at a lower intensity or slower repetition rate (as they do before they "warm up"), they would reduce their predation risk. This finding leads to the obvious suggestion that sexual selection is responsible for males behaving in this more risky manner.

In chapter 5 I showed that, like female túngara frogs, frog-eating bats also are attracted preferentially to calls with chucks. Again, this phenomenon needs to be interpreted in the context of the two opposing selective forces. Predation seems important in selection against maximum-call attractiveness. However, female choice also must be a powerful selective force, since it is the only explanation for why males produce chucks, rather than restricting their vocal repertoire to whines.

Obviously, the amount of calling by males throughout the night influenced their mating success. Also we noted that bats influence a frog's behavior in more subtle ways than by killing him. Tuttle, our field assistant Cindy Taft, and I often heard frog choruses fall silent as we saw a *T. cirrhosus* approach the breeding site. Therefore, we attempted to test the hypothesis that the mere presence of a *T. cirrhosus* influenced the amount of sexual display by male frogs. We were also interested in how túngara frogs might detect the presence of bats.

On six nights in March and April 1980, we used a night vision scope to record the precise times of visits by *T. cirrhosus* to Weir Pond on Lutz Stream, Barro Colorado Island. We also continually tape-recorded the *P. pustulosus* chorus. The tape recording was later played through a chart level recorder, giving us a readout of the number of calls throughout the period of observation. On the chart recording we then superimposed the time of bat visits, giving us a clear picture of how the presence of bats and chorusing behavior of the frogs were related (fig. 8.1). Since we assumed that the frogs relied primarily on vision to detect bats, we expected that the ambient light level would affect their perceptive ability. Therefore, we distributed our observations among nights with either a full moon or no moon.

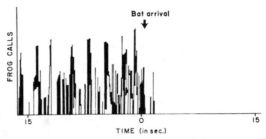

Figure 8.1 Chart recording made at Weir Pond showing typical response of chorusing túngara frogs to a bat arrival on a full-moon night. Vertical lines represent frog calls, the height of the lines represents the intensity of the call and is merely a function of the distance from the frog to the microphone (from Tuttle, Taft, and Ryan 1982).

We tested the hypothesis that the presence of bats influenced frog chorus behavior by comparing the number of calls produced 15 seconds before and after the bat arrived. The null hypothesis of no effect predicted that the number of calls produced in this 30-second period should be distributed randomly between the two 15-seconds periods. This was certainly not the case on full-moon nights, or on clear nights with no visible moon. A much smaller proportion of the calls during the 30-second period was produced after the bat arrived (fig. 8.2). The statistical analysis showed that on these nights, the 15-second period before the bat arrived had significantly more calls than the 15-second period after the bat arrived (table 8.2), thus supporting the graphical interpretation of the results (fig. 8.2). This was not the case on cloudy nights with no moon. On these nights the presence of the bats had little or no effect on chorus behavior; frogs tended to produce as many calls after a bat's arrival as before it. That it is too dark to see your hand in front of your face is a literal truth in the middle of the jungle on a cloudy, moonless

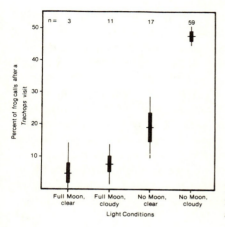

Figure 8.2 The proportion of calls made by frogs in the 15-second period following a bat arrival on nights with different light conditions.

Table 8.2 Mean Number of *Physalaemus pustulosus* Calls Produced in 15-second Periods before and after Arrival of *Trachops cirrhosus*

Light Conditions	N	Z statistic	P	Mean Number of Calls Emitted	
				Before	After
Full moon, clear	3	—	—	38.0	1.3
Full moon, cloudy	11	−2.93	0.002	27.9	2.5
No moon, clear	17	−3.39	< 0.001	31.1	9.2
No moon, cloudy	59	−0.98	0.164	53.2	50.8
All conditions combined	90	−4.36	< 0.001	45.4	35.4

Note: A Wilcoxon signed-ranks test was used to test the null hypothesis that there were likely to be as many calls produced in the period before the bat arrival as after the bat arrival (from Tuttle, Taft, and Ryan 1982).

night. The frogs likely did not respond to the presence of the bats because they did not see them.

These results demonstrate that the influence of bats on male túngara frogs is not an all-or-none proposition, that is, death versus survival. Bats can also influence the frogs in more subtle ways: merely by their presence, the bats influence the amount of calling by frogs. These results also suggest that frogs detect bats visually, and that on the darkest nights they may be unable to detect bats at all.

Predator Detection and Whom to Avoid

We also investigated the influence of light level on chorusing under more controlled conditions, where we could manipulate the light level and demonstrate more conclusively that the frogs visually detect hunting bats. Most students of animal behavior are familiar with the experiments by Tinbergen and Lorenz on form discrimination in turkey chicks. They showed that when a model was moved to resemble a hawk, it elicited a fear response from nestling turkeys, but it failed to do so when it was moved in the opposite direction and resembled a long-necked duck or goose (this study is summarized in many texts of animal behavior, e.g., Hinde 1970). These experiments are some of the most widely cited in the field of ethology. Tuttle, Taft, and I were familiar with these experiments and realized that we had always wanted to fly models over animals and watch their responses. Now we had a good reason to do so: by using models of bats, we could stage a bat visit at will under different light conditions. We could also manipulate the apparent risk to the frog by changing the distance between the frog and the bat model. Although we were not interested in form discrimination per se, we owe an intellectual debt to Tinbergen and Lorenz for the experimental design.

We constructed a two-dimensional model of *T. cirrhosus*, and installed a wheel on top of the model so that we could roll it along a string that passed over Kodak Pond. Using an artificial light source, we manipulated light level from that of a full moon, clear night ("normal", up to 35 lux), to "total" darkness.

Under normal light conditions, chorusing ceased every time we flew the bat model over the pond. However, chorusing also stopped in the apparent absence of any disturbance. We refer to this as a spontaneous chorus shutdown. Although it seemed unlikely, we had to be sure that the flight of the bat model did not just happen to coincide with a spontaneous chorus shutdown; therefore, we used spontaneous chorus shutdowns as our experimental control. We measured the time from chorus shutdown to reinitiation of calling, after spontaneous chorus shutdowns and when chorusing ceased after a pass of the model. The null hypothesis, that the bat model had no effect, predicts that the duration of the cessation of chorusing after a spontaneous chorus shutdown and after a pass of the bat model should not be significantly different.

On four nights under normal light conditions, the frogs ceased calling for a longer period of time in response to the bat model than when they stopped calling spontaneously (table 8.3). Under conditions of total darkness, the frogs did not stop calling when the model flew overhead, but no difference was found in the duration of the chorus shutdown between the apparent response to the model and the control (table 8.3).

When the bat model passed over the pond it emitted a slight noise due to the roller gliding along the string. In order to determine whether this noise, as opposed to the model itself, was eliciting the frogs' response, we repeated the same experiments but we used a wingless model. Again, the frogs ceased chorusing in response to the wingless model, but the duration of the chorus shutdown was not significantly different from the controls (table 8.3). Although the differences are not statistically significant, they do appear to be

Table 8.3 Duration of Chorus Shutdown in Response to *Trachops cirrhosus* Model

Night	# Frogs Present (beginning-end)	Mean Control (N)		Mean Bat Model (N)		t_s	P
Control vs. *T. cirrhosus* model under "normal" light conditions (0.17 to 0.35 lux)							
24 Mar.	24–22	25.5	(52)	112.2	(5)	1.67	< .0005
27 Mar.	30–32	33.7	(29)	353.6	(5)	3.25	< .0005
4 Apr.	29–26	25.5	(47)	193.2	(5)	3.65	< .0005
14 Apr.	17–23	27.7	(50)	201.6	(5)	3.66	< .0005
Control vs. *T. cirrhosus* model in "total" darkness (< < 0.17 lux)							
18 May	30–18	23.4	(19)	27.6	(5)	U = 52	NS
Control vs. wingless *T. cirrhosus* model (light level approximately 35 lux)							
11 May	39–42	33.4	(19)	52.6	(5)	1.38	NS

Source: Tuttle, Taft, and Ryan 1982.
Notes: Chorus shutdown, given in seconds, is defined as the time from the cessation of calling to the resumption of calling. Spontaneous shutdowns are control; other shutdowns occur after pass by model of *Trachops cirrhosus*.

large. While possibly the sounds elicited a slight response from the frogs, it is also likely, as I will discuss later, that the visual cues from even a wingless model elicited some response from the frogs.

The results of these experiments further support the conclusions of the field study: the presence of *T. cirrhosus* influences the behavior of the chorus, and the frogs use visual cues to detect the bat's presence. These conclusions do not suggest that the frogs do not respond to other cues provided by the bat as well. For example, we think that the frogs might hear the beating of the bat's wings, but we doubt that the frogs are able to hear the bat's echolocation calls, as some moths do (Roeder 1962). Although calling males are more likely to attract females, one can easily understand why the frogs should stop calling when a bat flies overhead. However, there are over 30 species of bats on Barro Colorado Island, and most of them pose no threat to the frogs. In fact, small insectivorous bats are relatively abundant at Weir Pond. If a male túngara frog stopped calling whenever he saw any of these bats, he would have precious little time available to advertise for mates. We wanted to know whether these small insectivorous bats elicited a repsonse from the frogs similar to the response elicited by *T. cirrhosus*. Therefore, we duplicated the experiments just described, but the same chorus was exposed to a model of *T. cirrhosus* (80 millimeters in body length) and a model of the small insectivorous bat, *Micronycteris megalotis* (60 millimeters).

The model of the smaller bat also elicited cessation of chorusing that was significantly longer in duration than the control. However, the chorus shutdown in response to the *T. cirrhosus* model was significantly longer in duration than the response to the *M. megalotis* model (table 8.4). Therefore, the presence of some bats that are of no threat to the frogs decreases the amount of calling by the frogs, but the frogs show a much greater response to the presence of *T. cirrhosus* than they do to the smaller insectivorous bat.

This conclusion need not imply that the difference in response is due to the ability of the frogs to identify different species of bats. Instead, their response may be due to differences in the size of the bats relative to the position of the frog. We suspect that we could elicit an even greater response from the frogs if we flew the top of a garbage can over their heads. This sensitivity to object size, however, does result in the frogs tending to show a greater response to bats that are dangerous.

Table 8.4 Duration of Chorus Shutdown in Response to *Trachops cirrhosus* and *Micronycteris megalotis* models

Shutdown Duration	N	Shutdown Duration	N	P
Control vs. *M. megalotis*				
29.4	(42)	54.9	(10)	< .005
M. megalotis vs. *T. Cirrhosus*				
54.9	(10)	189.4	(10)	< .005

We noticed that the response of the frogs to the bat model was not an all-or-none response, i.e., call versus do-not-call. Instead, the frogs exhibited a range of behaviors that varied in apparent evasiveness. The least extreme behavior would be for a frog to continue calling. All frogs stopped calling in response to bat models, but some frogs kept their lungs inflated and continued to float on the surface of the water, while other frogs deflated their lungs, keeping much of their bodies on the water's surface, or ducked down with only their snouts protruding. The most evasive behavior occurred when frogs dove under the water and swam away.

Does the bat actually influence the type of evasive behavior utilized by male frogs? If these behaviors are more or less evasive, as we assume, one should expect that the frogs exhibit the more evasive behaviors when the risk from predation is greater. We made an assumption that almost has to be true: the closer the bat is to the frog, the greater the probability that the frog will be captured by the bat. Therefore, we determined how the risk of predation influenced the frogs' behavior by passing the bat model at different heights over the pond, from one meter (on the high end of the slope, 0.91 meters on the low end; table 8.5) to 0.20 meters. Using a night vision scope, we watched several calling frogs before we flew the model and recorded the change in their behavior after the model passed. We assigned 1 for the least evasive behavior (continuing to call), and 5 for the most evasive behavior (diving; table 8.5).

As the model passed from 1.0 to 0.7 meters all of the males stopped calling, but they all remained floating on the water's surface with their lungs inflated—an average response of 2.0 (table 8.5). As the model was lowered, the frogs exhibited more evasive behaviors. At the lowest height, the response was the most evasive, 4.2, and 22 of the 36 frogs dove and swam away (table 8.5). A significant correlation was found between model height

Table 8.5 *Physalaemus pustulosus* Males Response to Model of *Trachops cirrhosus*

Model Height (meters)	Response Categories					N	Mean
	1	2	3	4	5		
0.11–0.20	0	7	1	6	22	36	4.2
0.21–0.30	0	5	1	1	3	10	3.2
0.31–0.40	0	6	3	2	1	12	2.8
0.41–0.50	0	26	2	2	2	32	2.4
0.51–0.60	0	23	1	1	0	25	2.1
0.61–0.70	0	13	0	0	0	13	2.0
0.71–0.80	0	17	0	0	0	17	2.0
0.81–0.90	0	3	0	0	0	3	2.0
0.91–1.00	0	2	0	0	0	0	2.0

Source: Tuttle, Taft, and Ryan 1982.

Note: Response categories are (1) continued calling, (2) stopped calling but remained inflated, (3) stopped calling, (4) deflated and lowered body until only top of head protruded above water, (5) dived.

and frog behavior: the lower the model, and presumably the greater the risk, the more evasive was the frogs' response (Spearman rank correlation, $r_s = -.96$, $P < .01$).

If a bat is in the area, why does a frog not use the most evasive behavior possible, since this would be the safest response? The answer is simple and obvious: a male that swam away from his calling site each time a bat was present would have little opportunity to attract mates—he might as well stay away from the breeding site altogether. But a male that adjusted his behavior to the risk of predation would have more time to spend advertising for mates, yet he could still reduce his risk of predation to some degree.

Not only does the magnitude of the predation risk explain why males gradually change their positions from floating on the water's surface to swimming away, but it also explains why males deflate their lungs as predation risk increases. A male with inflated lungs will be able to resume calling sooner after the predator passes, but a male with deflated lungs will be able to dive and swim away more quickly if approached by the predator, since an inflated male must first expel the air from his lungs before he can swim underwater. But energetic considerations as well might result in a male maintaining inflated lungs in the presence of a small predation risk. As shown in the last chapter (fig. 7.2), the incremental cost of producing a single call decreases as the calling rate increases. This partially results from the energy expended in reinflating the lungs by males that call infrequently. By maintaining lung inflation in the presence of bats, a male might not only reduce the amount of time needed to reinitiate calling, but he might also decrease the amount of energy expended for calling, relative to a male that always deflates his lungs when bats pass overhead. Even in a behavioral decision as seemingly trivial as whether or not to deflate the lungs, we see once again the forces of sexual selection and predation in opposition.

Why Do Males Join Aggregations?

If a calling male túngara frog reveals his presence to acoustically foraging predators such as *T. cirrhosus*, an aggregation of calling males would be even more conspicuous, and probably would be a favorite foraging site for these predators. To a predator, a chorus should be easier to locate and should contain more potential food items than a site with an isolated calling male. In addition, the competition among males for mates would seem to be greater in choruses. Why then, do male túngara frogs join choruses when advertising for mates?

Aggregations of sexually advertising males occur in a variety of animals, including insects (Alexander 1975), frogs (Wells 1977a), birds (Wiley 1974), and mammals (Bradbury and Verhencamp 1977). A number of suggestions have been made as to why males aggregate, but surprizingly few data have been collected. Tuttle, Taft, and I hoped to collect data to answer this question for túngara frogs by quantifying both the costs (as measured by predation risk) and benefits (as measured by mating success) obtained by males in choruses of different sizes.

Our attempt to measure rates of predation centered on the *P. pustulosus* chorus at Weir Pond. Not only was there predation by frog-eating bats at this site, but the South American bullfrog, *Leptodactylus pentadactylus*, an opossum, *Philander opossum*, and a crab, *Potamocarcinus richmondi*, also regularly ate túngara frogs at the pond.

The chorus size at Weir Pond varied greatly during April 1980. We determined the rate of predation on different nights, and compared the predation risks experienced by individual frogs in choruses of different sizes. Using a night vision scope, we recorded all of the predation events observed during periods of 1.5 to 3.0 hours on each of seven nights. Predation by the crab could not be observed with the night vision scope because it occurred underwater.

The bat *T. cirrhosus* was responsible for much more predation than either the bullfrog or the opossum. During a total of 14.3 hours of observation, the bats ate 95 frogs, an average of 6.6 frogs every hour (table 8.6). The rate of predation, which is the number of frogs taken per hour of observation, ranged from 2.5 to 12.1, but did not increase when chorus size was larger (Spearman rank correlation, $r_s = -0.28$, $P > .05$). Although larger choruses did not attract more frog-eating bats, we were interested in the risk of predation to an individual, not an entire chorus. We calculated the predation risk to an individual in the chorus by dividing the predation rate by the chorus size. As is shown in table 8.6, the risk of predation by bats was higher in smaller choruses ($r_s = -0.86$, $P < .05$).

The South American bullfrog commonly uses the same breeding sites as túngara frogs, but is much larger than túngara frogs (200 millimeters body length versus 30 millimeters; fig. 5.19), and has little trouble gulping down its unsuspecting neighbors. At least 2 and sometimes 3 bullfrogs were

Table 8.6 Predation Role and Risk as a Function of *Physalaemus pustulosus* Chorus Size

Date	Mean # of Frogs Present	Obs. Time (h)	Predation by									Total Predation Risk
			T. cirrhosus			*L. pentadactylus*			*P. opossum*			
			Obs.	Rate	Risk	Obs.	Rate	Risk	Obs.	Rate	Risk	
4/17	425	1.25	6	4.8	.011	1	0.8	.002	1	0.8	.002	.015
4/18	274	2.47	30	12.2	.044	1	0.4	.001	0	0.0	.000	.045
4/20	44	1.79	14	7.8	.178	1	0.6	.013	0	0.0	.000	.191
4/21	50	2.59	16	6.2	.124	0	0.0	.000	5	1.9	.039	.163
4/22	340	1.98	5	2.5	.007	11	5.5	.016	5	2.5	.007	.030
4/23	330	2.38	17	7.1	.022	1	0.4	.001	7	2.9	.009	.032
4/27	130	1.85	7	3.8	.029	0	0.0	.000	0	0.0	.000	.029

Source: Ryan, Tuttle, and Taft 1981.
Note: Predation rate is frogs eaten/hour observation; predation risk is predation rate/chorus size for each predator.

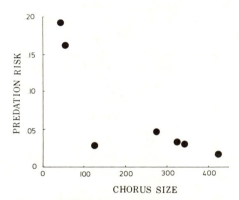

Figure 8.3 Predation risk (equals predation rate divided by chorus size) for individual frogs in choruses of different sizes.

present during our observations, and we saw them eat 15 túngara frogs, an average of 1.1 frogs per hour. As with predation by bats, neither the rate of predation on the chorus ($r_s = 0.49$, $P > .05$) nor the risk to individuals in the chorus ($r_s = 0.20$, $P > .05$) from bullfrogs showed a statistically significant increase with chorus size (table 8.6). Opossums visited the pond on four of the seven nights of observation, and they ate 18 frogs, an average of 1.3 frogs per hour. With this predator as well, neither the predation rate ($r_s = 0.49$, $P > .05$) nor the predation risk ($r_s = 0.20$, $P > .05$) was significantly correlated with chorus size (table 8.6). The data combined for all three predators reveal that larger choruses do not experience a signficantly higher rate of predation ($r_s = 0.11$, $P > .05$), and the larger the chorus, the lower the risk of predation for any individual in that chorus ($r_s = -0.78$, $P < .05$; table 8.6; fig. 8.3).

Males advertise to attract mates. Although they might experience a lower risk of predation in larger choruses, we must ask: at what price? Is there a trade-off between the safety of a large chorus on the one hand, but increased competition for mates on the other? As should be clear from chapter 4, accurate data on male mating success can only be obtained from continual monitoring of a population of individually marked males. These data are not available for the population at Weir Pond. However, the data I collected on male mating success at Kodak Pond allowed us to determine the relationship between male mating success and chorus size in general.

In both 1978 and 1979 I found a significant positive correlation between chorus size (the number of males) and the number of females that were attracted to the chorus and mated (1978: $r = 0.67$; $P < .01$, $N = 48$; 1979: $r = 0.73$, $P < .01$, $N = 152$). Because the data from 1979 were more extensive, we examined the number of females that were attracted to the chorus as a function of chorus size (fig. 8.4). Although the data points are scattered, the results of the linear regression analysis are consistent with the correlation analysis and show that chorus size does explain a significant amount of the variation in the number of females among nights (table 8.7).

Although the number of available females increases with chorus size, this does not allow us to determine how the average male's mating success changes with chorus size. Therefore, we fitted the data with a second-order

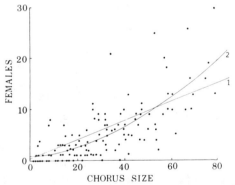

Figure 8.4 Number of females versus chorus size at Kodak Pond in 1979 with the resulting linear (1) and second-order (2) regressions plotted (from Ryan, Tuttle, and Taft 1981).

Table 8.7 Analysis of Variance for the Linear and Second-order Regression of the Number of Mated Females on Chorus Size

	df	ss	F
Linear regression	1	2004.9	162.4
Residual	151	1790.2	
Linear regression equation: $Y = -0.79 + 0.20 X$			
Second-order regression	2	2117.7	90.9
Residual	150	1677.4	
Second-order regression equation: $Y = -0.82 + 0.05 X + 0.002 X^2$			

regression and found that this regression explains significantly more of the variation in the data than does the linear regression ($F = 9.7, P < .005$); the second-order regression shows that the number of females tends to increase at a faster rate than does the chorus size (fig. 8.4). This trend is also apparent from the positive B_2 term in the second-order regression equation (table 8.7). Since the ratio of males to females increases with chorus size, so does the average male mating success. Although a visual inspection of figure 8.4 does not reveal a striking trend, the analysis clearly demonstrates that a male does not trade off mating success for safety from predators by joining a larger chorus. If anything, he seems more likely to mate in these choruses. The relationship between the number of females and chorus size at Kodak Pond involves much smaller choruses than those that were at Weir Pond. Although we doubt that the quantitative relationships are precisely the same at both sites, we see no reason to expect qualitative differences in the relationship.

In this study we observed predation on one species of frog at one chorus site. How general are these measures for *P. pustulosus* and for other anurans? Although I did not observe bats catching túngara frogs at Kodak Pond, on a number of occasions I did observe predation by the marine toad, *Bufo marinus* (see also Jaeger 1976), juvenile pond turtles, *Pseudemys scripta*, and the cat-eyed snake, *Leptodeira annulata*, which is a well-known frog predator (Duellman 1958). In the absence of a broad data base, my surmise is that túngara frogs experience a high level of predation, but also that considerable variation exists both in the frequency of predation and in the predators involved, and that much of this variation is due to the particular habitat in which the frogs breed.

P. pustulosus may not be unique in the high rate of predation it suffers. In a detailed study of the natural history of the gladiator frog, *Hyla rosenbergi*, in Panama, Kluge (1981) witnessed predation by two animals that commonly feed on túngara frogs—the South American bullfrog and the cat-eyed snake. He also noted that no resightings of permanently marked males occurred between 1978 and 1979; he suggested that high predation pressure places *H. rosenbergi* among the shortest lived anurans. Nor is predation restricted to tropical frogs. The long-term study of bullfrogs by Howard (1981) suggests that a higher predation rate on males partially explains why females in this species are larger than males: the females, on average, live longer than the males. Therefore, even though the data on predation that we collected are only derived from one species at one chorus, they suggest that predation might play an important role in the social organization of other species as well.

The selective forces responsible for the evolution of aggregations of sexually displaying males have been widely discussed. Hamilton (1971) suggested that reduced predation risk in larger aggregations might explain the centripetal tendencies of many species. Given a constant rate of predation, an individual is safer if he is part of a group. Our results are consistent with this concept, which is popularly known as the selfish-herd theory. However, this does not necessarily explain why females are attracted to larger choruses.

Several suggestions have been offered to explain why females are attracted preferentially to larger aggregations of sexually displaying males. We can dispose of two of these hypotheses with regard to túngara frogs. First, Alexander (1975) suggested that some aggregations of sexually displaying males could be classified as resource-based aggregations. By displaying at resources utilized by females for reproduction, males would be sure of encountering females. Male túngara frogs do advertise from resources needed by the female for reproduction—small bodies of standing water. But if the resources were the only factor responsible for the spacing of males and the attraction of females, then males should be randomly spaced over that resource. This is clearly not the case with túngara frogs, since males do not call from some bodies of water that appear to be ideal breeding sites. I also have some experimental evidence that tends to support this suggestion.

Table 8.8 Number of Male *Physalaemus pustulosus* Calling from Each Bucket

Date	Bucket #1	Bucket #2
3/03/81	2*	1
3/04/81	1	2*
3/05/81	1*	0
3/08/81	1	1*
3/09/81	2*	0
3/10/81	2	0*
3/15/81	1*	0
5/18/81	2	4*
5/21/81	5*	3
5/23/81	0	1*

Note: * represents the bucket that had the simulated chorus.

On ten nights I placed two small buckets (⅓ meter in diameter) in the ground about 10 meters apart, in an area where túngara frogs were known to breed but were not abundant. Both buckets were filled with water, and a speaker that continually broadcasted a túngara frog chorus was switched between buckets on successive nights. Very few males came to call from either of the experimental sites, but on eight of the ten nights more males called from the bucket with the artificial chorus, and on one night the same number of frogs came to each chorus (table 8.8). Although these differences are not statistically significant (two-tailed exact binomial probability test, $P = .07$), they do suggest that the presence of other males, and not merely an appropriate resource, might be responsible for the aggregation of calling males. Clearly, more data are needed to evaluate this hypothesis.

A second suggestion as to why males join choruses is that these males will have a greater chance of mating because larger choruses attract more females, since the pooled acoustic stimulus of a larger chorus travels over a greater distance than does the acoustic stimulus of a smaller chorus (e.g., Wells 1977a). Bradbury (1981) has shown that the effect of the pooled stimulus alone predicts only that larger aggregations will attract absolutely more females, not proportionately more females, than a smaller aggregation. He argues that the number of females attracted to a chorus is dependent on the active space of the chorus, that is, the area over which a female can hear the chorus. If males produced calls of identical frequencies that were perfectly in phase, the active space and hence the number of females attracted would increase linearly with chorus size. The proportion of females would remain constant as chorus size increased, and so would the probability of a male mating. Since males do not call perfectly in tune and phase, the pooled-stimulus argument predicts that proportionately fewer females should visit larger choruses. As we have seen, this is not the case in *P. pustulosus*.

The hypotheses that females are attracted to larger choruses either because these choruses are formed at resources needed by the females for reproduction, or because larger choruses attract females over a greater distance, do not explain why proportionately more female túngara frogs are found at larger choruses. Two other hypotheses are that females experience a lower predation risk when they mate at larger aggregations (e.g., Wittenberger 1978), and that they prefer larger choruses because these contain more males from which to choose a mate (e.g., Emlen and Oring 1977; Bradbury 1981). Both hypotheses have credence in this case. Some predators, such as bats and opossums, use acoustic cues while hunting, but they probably use other senses as well, including vision. Other predators, such as snakes and land crabs, may depend on vision or tactile cues to locate prey. Although females probably have a much lower risk of predation than calling males, they too might be vulnerable to predators attracted to the general area by the acoustic stimulus of the chorus, or to predators that hunt by senses other than hearing. As with males, females are probably safer in larger choruses.

Clearly female túngara frogs exhibit mate preference among conspecific males. This finding is consistent with, but does not prove, the hypothesis that females are attracted to larger choruses because they afford more opportunities to choose mates. But one must remember that these data for túngara frogs compare choruses of different sizes on different nights at the same breeding site. Perhaps females do not sample and compare choruses of different sizes. If males and females had different thresholds at which environmental conditions triggered migration to the breeding site, the proportional increase of females at large choruses still might result. The data that are currently available do not permit us to distinguish between these alternative hypotheses, and in fact, they need not be mutually exclusive.

These results suggest that some adaptive advantage exists to the chorusing behavior of male *P. pustulosus*. Not only do males in larger choruses have a lower risk of predation, but on average they also have a better chance of mating. Although it is not clear which, if any, of these advantages was the selective force responsible for the evolutionary origin of chorusing behavior in túngara frogs, both advantages are probably important for its evolutionary maintenance.

Summary of Predation Risk

Bats, opossums, and other frogs eat túngara frogs, locating them by orienting to the calls of males. By attempting to attract mates, males considerably reduce their chances of survival. Also, changes in a male's vocal behavior that would increase his ability to attract females, such as calling more intensely or producing calls with chucks, further increase his predation risk. Under certain conditions the frogs are able to detect an approaching bat. Although they show a greater response to the frog-eating bat, the frogs also exhibit a lesser response to bats that pose no apparent threat to their survival. Therefore, the mere presence of a bat influences the amount of calling by males, and the distance from the bat to the frog determines the specific type of antipredator behavior that is employed. Interactions between frog-eating bats and túngara frogs highlight the opposing effects of selective forces generated by sexual selection and predation.

Although predators are attracted to the calls of túngara frogs, larger aggregations of calling males do not result in a higher risk of predation to individuals in the chorus. This selfish-herd effect might be one of the reasons why males prefer to call in aggregations instead of in isolation. Males also do not sacrifice mating success for safety in larger choruses. The proportion of females attracted to choruses, and thus the average mating success of males in the choruses, does not decrease as a function of chorus size; in fact, it tends to increase.

Several hypotheses are consistent with the observation that proportionately more females are present at larger choruses: (1) females as well as males prefer larger choruses because predation risk is reduced by the selfish-herd effect; (2) females prefer larger choruses because they provide more opportunities to choose mates; (3) females do not actually sample choruses and prefer the larger ones, but the physiological mechanism that triggers migration to the breeding site is such that proportionately more females are active on nights when there are larger choruses. These hypotheses are not necessarily mutually exclusive, and the data needed to evaluate each alternative are not available.

Reproductive Costs and Reproductive Behavior

As discussed earlier, the concept of the cost of reproduction is important to theories of sexual selection and the evolution of life-history patterns. In *P.*

pustulosus, the amount of energy expenditure and predation risk involved in reproduction has implications for sexual size dimorphism, the age and size at first reproduction, and the lack of alternative mating strategies in this species. These are considered in turn.

Sexual Size Dimorphism

In many polygamous species in which large size affords a mating advantage for males, the males are the larger of the two sexes (e.g., pinnipeds; Bartholomew 1970). Similarly, male túngara frogs might be expected to be larger than females, since larger males have a higher probability of mating. However, selection for large size in females should occur as well, since clutch size increases with female body size (chap. 6). This implies that selection should tend to maximize body size in both sexes, and as a result, the sexes should be the same size. However, such is not the case in *P. pustulosus*—females are larger than males. In fact, larger males enjoy greater mating success in a variety of anuran species, but in 90% of the 589 species surveyed by Shine (1979), females are larger than males. Sexual differences in age at first reproduction, predation risk (Howard 1981), and expenditure and acquisition of energy (Woolbright 1983) have all been proposed to explain why females are larger than males in some species of frogs.

If one sex expends a larger proportion of energy for reproduction than the other, that sex might be smaller because it has less energy available for growth. In *P. pustulosus,* females expend much more energy for reproduction than do males, but they are also the larger sex; therefore, sexual differences in the absolute amount of energy expended for reproduction do not explain the difference in body size between male and female túngara frogs. Woolbright (1983) points out that females might have a greater intake of energy since they have more time to forage than males, which are engaged in time-consuming reproductive behaviors. He suggested that this might explain why females often are the larger sex; females can expend more energy for reproduction yet still have more energy for growth than males, if they have sufficiently more time available for foraging. This explanation might apply to *P. pustulosus,* since almost one-half of the males collected at a breeding site in Gamboa had empty stomachs (table 7.8). Males probably do not feed while they are at the chorus. However, increased foraging time for females could also lead to increased predation risk away from the breeding site, which would tend to decrease the average size of females. To fully evaluate the role of energetics in sexual size dimorphism, we need to know not only the amount of energy expended in reproduction by each sex, but also the total amount of energy acquired and the risk of predation during foraging. Although not all these data are available, it is probably safe to conclude that the large difference between the sexes in energy expended for reproduction does not explain why female túngara frogs are larger than males.

Túngara frogs continue to grow after reaching sexual maturity (fig. 6.1), as do many other vertebrate ectotherms. Therefore, the size distribution of each sex is related to its age distribution. We have demonstrated that a great deal of predation on *P. pustulosus* occurs at the breeding site. The males are more likely to be located by acoustically foraging predators because they advertise their presence and their location. Males are also exposed to more predation because they are present at the breeding site on more nights than the females. Even on the night a female visits the breeding site, a given male remains there longer and is more conspicuous. Unless females forage more than males, mortality away from the breeding site must be assumed to be the same for both sexes. The sexual size dimorphism in *P. pustulosus* is likely due largely to higher survival rates of females—they live longer, so they are larger. This is also true in bullfrogs (Howard 1981), and I suspect that it might be the case for a number of species of frogs. Another possible factor is age at first reproduction, which also skews the size distribution towards larger females at the breeding site in bullfrogs, since females begin to breed at a later age (Howard 1981). And female túngara frogs also begin breeding at a larger size than males (fig. 6.5), which might indicate that they too begin to breed at a later age.

When to Reproduce?

The evolution of life-history patterns, like other aspects of an organism's biology, is influenced to some extent by natural selection (e.g., Stearns 1976). One life-history parameter that has been investigated extensively is the age at which an individual first reproduces (e.g., Pianka and Parker 1975). An obvious prediction is that a male should not invest in reproductive behavior unless he has some probability of mating; the associated costs of reproduction, as well as the probability of mating, should affect the optimal age at first reproduction.

In the túngara frog, the absolute ages of reproductively active males are not known, but size provides a rough estimate of relative age. Thus given the observed costs and benefits of reproductive behavior, we can ask whether males begin to reproduce at a size that is predicted by life-history theory. The probability of mating increases with male size, and the smallest males that visited the breeding site—24 millimeters—did not succeed in mating at all (fig. 4.6). Although males smaller than 24 millimeters might have some chance of mating, it appears to be very small. Therefore, males should postpone reproduction until some time after they are 24 millimeters long. This seems to be the case, since in the laboratory males are "capable of reproducing [at] . . . one-half to two-thirds the average length of fully grown adults" (Davidson and Hough 1969)—that is, at approximately 20 millimeters.

In the absence of associated costs of reproduction, male túngara frogs should begin breeding at about 25 millimeters. However, the risk of predation during reproduction should induce males to postpone reproduction: a male should not take the chance of being killed unless he has a good chance of mating. A computer simulation makes some predictions about when a male should begin to breed at a size that will maximize his lifetime mating success. For simplicity, this model does not take into account the consideration that the earlier the offspring are produced, the sooner they themselves will reproduce, so that offspring produced earlier are "worth more" in terms of potential reproductive output (Cole 1954). The probability of a male mating at each size is taken from the 1979 data on male mating success (fig. 4.6), and the time spent by a male at each size is estimated from the average growth rate of males of each size (fig. 6.1). We assume that a male spends 17% of the nights at the breeding site, based on the frequency of attendance by marked males in the 1979 study.

As stated earlier, if no risk of predation existed at the breeding site, males should begin to reproduce at 25 millimeters. However, if the average probability of being killed at the breeding site is 1 out of 100 (0.01), then males should postpone reproduction until they are 31 millimeters long (table 8.9). When the predation risk per night increases to 0.02 and greater, then males should only assume the risk of advertising for females when their chance of mating is greatest, at a body size of at least 34 millimeters (table 8.9).

Table 8.9 Optimal Size at First Reproduction for Given Predation Risk at and away from Breeding Site for Male *Physalaemus pustulosus*

Predation Away from Breeding Site	Predation Risk at Breeding Site										
	0	.01	.02	.03	.04	.05	.06	.07	.08	.09	.10
.00	25	31	34	34	34	34	34	34	34	34	34
.01	25	29	31	31	34	34	34	34	34	34	34
.02	25	29	31	31	34	34	34	34	34	34	34
.03	25	27	31	31	34	34	34	34	34	34	34
.04	25	27	29	31	34	34	34	34	34	34	34
.05	25	27	29	31	34	34	34	34	34	34	34
.06	25	25	25	25	25	25	25	25	25	25	25

The predation risks used in the computer simulation are at the lower end of values observed at Weir Pond, which are based on an hourly and not a nightly basis (table 8.9). Yet the model predicts that males should postpone reproduction until much later than they do. At least three possible explanations for this discrepancy can be found: (1) The predation rate at Kodak Pond is much lower than it is at Weir Pond. If it is close to zero, then the observed behavior is consistent with the predictions of the model. (2) The predictions are correct but selection, for any of several reasons, has not resulted in an evolutionary response in the reproductive behavior of males. (3) Significant mortality occurs away from the breeding site. This would select for an earlier age, and hence smaller size, at first reproduction.

The first possible explanation seems very unlikely. I assume that the predation risk for túngara frogs does vary among different breeding sites, and bats were not observed at Kodak Pond. But my frequent observations of predation by cat-eyed snakes, marine toads, and pond turtles at Kodak Pond suggest that even if the predation risk is lower there than at Weir Pond, it is still substantially above zero. The second explanation can confound the predictions of any life-history model (e.g., Stearns 1976, 1980), or any prediction based upon maximization of fitness (e.g., Lewontin 1977). It is certainly possible in this case as well, but unfortunately, this explanation cannot be tested experimentally.

The third possibility is an important one. Clearly, mortality occurs away from the breeding site, and this drastically changes the predictions of the model. No data are available on nonbreeding mortality, and estimating it from mark-recapture data can be accomplished only under several very unrealistic assumptions. Even without these data, it is still interesting to enter a range of values for nonbreeding mortality into the model. Regardless of the extent of nonbreeding mortality, if no predation occurs at the breeding site, males should breed as early as possible (table 8.9). As nonbreeding mortality increases, and when the chance of being killed away from the breeding site is 6 out of 100 (0.06) on any day, males should attempt to reproduce as early as possible. Predation risk at the breeding site for *P. pustulosus* is high, as it probably is for other tropical anurans (e.g., Kluge 1981). Given the observed size distribution of reproductively active males, we must assume either that males also have a relatively high rate of nonbreeding mortality, or that selection has had little or no influence on age at first reproduction in this species, even though it has in other anurans (e.g., wood frogs; Berven 1982).

Alternative Male Reproductive Behaviors

Larger males are more likely to be selected as mates by females, and calling males, whether large or small, have a higher risk of predation than when they do not call. In some species of frogs, a subset of males exhibit noncalling reproductive behaviors known as "satellite" or "parasitic mating strategies." These males are present at the breeding site but do not call; instead, they attempt to mate with females attracted by the calling males (Perrill, Gerhardt, and Daniels 1978). In species in which larger males have a mating advantage, smaller males that adopt these alternative reproductive behaviors probably enjoy a greater reproductive success than they would if they exhibited "normal" reproductive behaviors (e.g., Howard 1978a). Because a large-male mating advantage occurs in *P. pustulosus*, and because of the high cost associated with sexual displays (i.e., predation and energy expenditure), an alternative mode of male reproductive behavior might be predicted to occur. But it does not.

Although on any given night some males call more than others, a high correlation is found between the amount of time a male is present at the site and the amount he calls. This pattern would not be expected if a subpopulation of males adopted noncalling reproductive behaviors. If we assume that males have the behavioral flexibility to adopt noncalling behaviors and that males are acting in a manner that increases their mating success, then possibly the decreased risk of predation and the savings of energy expended in calling fails to offset the reduced mating success that would be experienced by a noncalling male. We need to consider the degree to which predation risk and energy expenditure would be reduced, and whether or not a male would have any chance of mating if he did not call.

Calling males attract some predators, but as noted above, these predators probably use additional cues to search for frogs, especially since the frogs usually stop calling when predators arrive. The closer a noncalling male is located to a calling male, the more likely he will be able to intercept a female en route to a calling male, but he will also be more likely to be located by a predator initially attracted by the calling male. Thus for a noncalling male, the risk of predation and the probability of intercepting a female should both increase with decreasing distance to a calling male. Perhaps to have any significant chance of intercepting a female, a noncalling male would have to be so close to a calling male that the reduced predation risk of silence would be minimal.

Some authors (e.g., Perrill, Gerhardt, and Daniels 1982; Gerhardt 1983) have suggested that a noncalling male decreases the amount of energy expended for reproduction. Our results show that this is true for túngara frogs, although even for a noncalling male some increase occurs in metabolic rate, relative to nighttime resting rate, that apparently results from the acoustic stimulation of the chorus. Whether decreased energy expenditure would significantly increase a male's future mating success is not clear: male *P. pustulosus* use far less energy for reproduction than the apparent potential for this species, based on the strikingly larger energetic expenditures of females. But the facts that males do not seem to feed while they are at the chorus, and that while calling they expend energy at a rate 1.7 times higher than if they were at the breeding site but not calling, might explain why males are only present at the breeding site for a small proportion of the time (17%) they are known to be alive. Energy conservation might enable a noncalling male to increase the number of nights he visits the breeding site. The number of nights at the breeding site is the most important factor affecting a male's mating success, but only if the male calls. Whether or not a noncalling male would experience substantially increased mating success with increased time at the pond is not possible to predict.

Of course, noncalling behaviors are only a viable reproductive option if females can be obtained by males that do not call. We cannot assume that females are simply projectiles that passively move through the chorus waiting to be seized by any properly motivated male. This does not appear to be true for túngara frogs (chap. 3), and it probably is not true for many other frogs, even species with explosive breeding seasons. For example, female *Bufo cognatus* are known to exhibit a variety of behaviors that result in their active avoidance of noncalling males (Sullivan 1983), and similar behaviors have been reported for female *B. americanus* (Waldman 1980). And when these females are intercepted by noncalling males, they can effectively displace the unwanted male (see also Davies and Halliday 1977). Perhaps the preference of females for calling males, and their potential ability to avoid noncalling males, has made noncalling reproductive behaviors a pointless option for túngara frogs. In this species, female preference for calling males must be an important selective force, because it results in males performing a suite of behaviors that reduce their probability of survival. Not only do males advertise their presence to a variety of predators, but they produce types of calls that further increase their chance of being killed—all due to the influence of female choice.

Summary of Reproductive Costs and Reproductive Behavior

Males incur a high cost for engaging in reproductive behaviors due to the increased risk of predation; both sexes, especially females, incur a cost due to the considerable amounts of energy that they expend in the effort to reproduce. These costs of reproduction have implications for the observed patterns of sexual size dimorphism, age and size at first reproduction, and male reproductive behavior in túngara frogs.

Despite the fact that a large-male mating advantage is present in túngara frogs, females are the larger sex. This apparent contradiction occurs in many other frog species as well. Because females expend considerably more energy for reproduction than do males, sexual differences in energy expenditure do not explain the observed sexual size dimorphism. Females probably are larger because they live longer, and the lower survival rates of males result from the cost of predation associated with calling behavior.

Males seem to postpone reproduction until they are a size at which some possibility of mating exists. Given the predation risk associated with calling, the observed size distribution of reproductively active males suggests either that significant mortality occurs away from the breeding site or that selection has not influenced this aspect of the life history of túngara frogs.

In spite of the costs of predation and energy expenditure associated with calling, male túngara frogs do not exhibit noncalling patterns of reproductive behavior, as do some other frog species. Three factors may contribute to this. First, after a predator is attracted to a calling male it will use other cues, such as vision or echolocation, to locate its prey. For a noncalling frog, both the probability of intercepting a female and the risk of predation will increase as he moves closer to a calling male. Perhaps to have any chance of mating, a noncalling male would have to be so close to a calling male that the reduced risk of predation, relative to the risk experienced if he were calling, would be minimal. Second, males conserve energy by not calling constantly, and this might increase the total number of nights a male visits the pond. For calling males, the likelihood of mating success would be significantly enhanced, but it is not known if this would be true for noncalling males as well. Unless a noncalling male has some chance of mating, increasing the amount of time at the breeding site only increases his predation risk. Finally, we must consider whether the behavior of female túngara frogs excludes noncalling as a viable option for reproductive behavior. Studies of other species of frogs show that females have a suite of behaviors that enables them to avoid mating with noncalling males. Possibly túngara frog females also possess this potential.

9 Conclusions, Speculations, and Suggestions

The five main questions addressed in this study are: (1) Does female choice influence male mating success? (2) Which males do females select as mates? (3) On what characteristics do females base this choice? (4) What selective advantages, if any, are obtained by females choosing certain males? and, (5) How do the associated costs of reproduction—predation, and energy expenditure—influence the species' mating system? In this final chapter, I briefly review the answers to these main questions and to other questions that arose during the study. I then speculate on the evolutionary significance of these results, and suggest where the gaps of information in this and in other studies are, and how these gaps might be filled.

Male túngara frogs aggregate at breeding sites and conspicuously advertise their presence to females. The females enter the breeding site, move among the calling males with little or no interference, and eventually initiate mating with a calling male. Clearly, the behavior of females influences, if not determines, which males at the breeding site will mate. Túngara frogs are not unusual in exhibiting this type of female-choice behavior. Females choose mates in many species of frogs with prolonged breeding seasons. Nor is this phenomenon restricted to frogs; the behavior exhibited by females in species of birds in which males display on leks is only one analogous example of female choice.

In this study, detailed behavioral observations documented the fact that females behaved in a way that could be interpreted as choice. However, in studies of other mating systems, such as some of the studies of toads, female choice has been inferred from data that demonstrate nonrandom mating among males. These studies have given little or no attention to the mechanisms that result in this nonrandom mating. In studies of anuran species where scramble competition among males is prevalent, it is especially important that detailed behavioral observations of both sexes prior to mating are conducted. Only these observations will reveal whether females have any opportunity to choose mates. Although experimental techniques, such as playback experiments, can rigorously test hypotheses about the mechanisms of female mate choice, these experiments are meaningful only if they are derived from a solid knowledge of how the animals behave in nature. And this knowledge can only come from watching the animals—and watching them for a long time.

When I compared the sizes of mated and unmated males at the breeding site, for the entire study as well as on a nightly basis, I found that larger males were more likely to mate. This observation is not unusual. Large male size affords a mating advantage in a variety of animals. In frogs this occurs in species with both explosive breeding seasons and prolonged breeding seasons; in the latter situation, females choose larger males as mates. Exactly how they do it is usually not investigated.

Male túngara frogs do not defend territories, so cues available for mate choice to females of some species, such as bullfrogs, are not available to female túngara frogs. Males do vary considerably in their reproductive behaviors, and there is little doubt that much of this variation influences female choice of mates. But does any of this behavioral variation among males explain the size-biased male mating success observed at the breeding site? Differences among males in the amount of calling, the intensity of the call, the location of the calling site, and various temporal and spectral features of the call do not vary predictably with male size. These factors might influence which males are chosen as mates, but they do not explain why females tend to select larger males. However, one factor does vary predictably with male size—the fundamental frequency of the chuck portion of the mating call; larger males have calls with lower frequency chucks. In playback experiments females were attracted preferentially to chucks with lower frequencies, which explains, at least in part, why larger males are more likely to mate.

Sexual displays often contain information that allows females to identify males as conspecifics, but many, if not all, species show some variation in sexual displays. I suggest that due to a preoccupation with species-isolating mechanisms, intraspecific variation in sexual displays and the significance of this variation in influencing differential male mating success has been largely ignored. It is important to consider not only the stereotypy of these displays but also the variation. In many species, females will likely be attracted to some courtship variants instead of others, and investigation of this sort of phenomenon will be necessary for a complete understanding of how sexual displays evolve.

Túngara frogs are unusual in that the sexual display contains two components with very different functions: the whine is sufficient and necessary for species recognition, while the chuck does not contribute to species recognition and seems to function in providing additional information (e.g., size of the male) to females selecting mates. Clearly, the chuck has evolved under the influence of sexual selection. Not only is the chuck per se a result of female choice, but a phylogenetic comparison demonstrates that due to the low frequencies of the chuck, túngara frogs have calls with relatively much lower frequencies for their size than closely related species. Outgroup comparisons suggest that the low frequency call is the derived state. Thus data from the field and from playback experiments document the action of selection through female choice, and a phylogenetic analysis suggests that an evolutionary response to this selection has occurred.

I think that the phylogenetic approach is a technique that has too often been neglected by modern behaviorists. Most studies in behavioral ecology address the potential adaptive significance and current utility of behavior. Often these studies do not even consider the question of how the behavior evolved. If we are truly interested in behavioral evolution we must combine approaches that address mechanism, utility, and history. And the only technique that allows us to test predictions about the direction and rate of behavioral evolution is the phylogenetic approach. Like all approaches, it has strong points and weak points (e.g., ideally, we should know something about the behavior of all the closely related species). However, I urge that this approach be incorporated into studies of sexual selection whenever possible, and especially that we make use of some of the formal techniques that have been developed by systematists to study character evolution (e.g., Wiley 1981).

Female túngara frogs select mates, and their preferential attraction to certain call characteristics results in their tendency to select larger males. To some extent, we know how these females choose larger males. Identifying the evolutionary forces responsible for the origin and maintenance of female choice in this species is more difficult. Several hypotheses have been formulated: (1) female preference for lower frequency calls is merely an epiphenomenon of the structure of the female's auditory system; (2) female choice has evolved because larger males have genotypes better adapted for survival and growth (females might be selecting the "best" genotypes or avoiding the "worst" genotypes); (3) female choice has evolved because larger males enhance the reproductive success of females; (4) female choice has evolved in concert with the male trait through the process of runaway sexual selection.

We have no data that allow us to reject the first two hypotheses. In fact, data to test the second hypotheses—a genetic advantage through natural selection—are rarely available for any species. In regard to the third hypothesis, the data available suggest that for most females, choice of larger males increases their reproductive success because it decreases the number of unfertilized eggs. The effect is small and its evolutionary significance is difficult to evaluate. The fourth hypothesis is difficult to test experimentally. However, the prediction tested by the phylogenetic analysis does suggest that female choice has influenced the sexual display of túngara frogs. This indicates the possibility that runaway sexual selection may be operating, because female choice has affected the display itself (the lower frequency of the chuck) rather than the phenotypic quality of the male (i.e., his size) with which the display correlates. Therefore, both a nongenetic advantage due to natural selection (hypothesis 3) and a genetic advantage due to sexual selection (hypothesis 4) might select for females that mate preferentially with larger males.

One important gap of information in almost all studies of sexual selection is the lack of data on the heritabilities of traits correlated with male fitness. This is a perplexing problem, and I suggest that even if the data were available they might be difficult to interpret. Because female choice is based on phenotypic variation, it will proceed in the absence of any genetic basis to the variation. As mutations (or other forces that increase genetic variability) occur and the heritability of a trait increases, evolutionary change should occur in that trait if it is subject to sexual selection; but this force will in turn deplete the heritability of the trait. Therefore, the evolutionary history of a male trait under sexual selection might be characterized by long periods of stasis interrupted by short periods of change. If the heritability of a trait is determined, we then know the potential for this trait to evolve in response to selection at that point in time. But does that tell us anything about the role of sexual selection in the evolutionary history of the trait (i.e., how it got where it is now), or does it merely tell us that we are examining the trait during a period of stasis or change? Amassing data on the heritabilities of male fitness traits is important, but will not solve all our problems; sexual selection studies should be supplemented by a phylogenetic approach to shed additional light on the evolution of sexual displays.

The costs associated with reproduction have had an important influence on the evolution of reproductive behavior in túngara frogs. The risk of predation incurred by calling males partially explains the variable complexity of the advertisement call, why males aggregate while advertising for females, and why males do not reproduce at the smallest size possible. The influence of predation on the reproductive behavior of túngara frogs highlights how sexual selection and natural selection can act in opposition, and especially how sexual selection can result in the evolution of traits that are maladaptive for survival. I hope the results of this study will demonstrate the need to consider the importance of predation in other animal communication systems, especially those involved in sexual selection.

Energy expenditures have a more subtle influence on reproductive behaviors than does predation. In particular, the energetic cost of calling should strongly influence the amount of time that a male spends at the breeding site. This is an important influence (and one not sufficiently investigated in this study), since the number of nights at the breeding site is the most important determinant of male mating success. A male's size does not influence how much time he spends at the breeding site, but a variety of other factors, such as the ability to locate and process food, must play an important role. No doubt there is a large stochastic component to finding food, but to some degree the energetic cost of calling will act as a filter, eliminating from (or reducing the amount of time at) the breeding site those males not capable of acquiring sufficient resources.

Túngara frogs have an unusually high dependence on aerobic metabolism relative to other frogs that have been studied. Although this dependence is correlated with foraging and type of locomotion, in túngara frogs a high aerobic capacity also enables frogs to support a variety of reproductive behaviors over fairly long periods of time. It would be interesting to know how aerobic capacity might evolve in response to selection on reproductive behaviors, as well as how aerobic capacity constrains the behavioral options available to a male, and whether differences in aerobic capacity among conspecific males have any influence on the ability of individual males to acquire mates.

Admittedly, this study of sexual selection and communication in túngara frogs has not answered all the question. My attempt to integrate investigations of the species' communication system, behavioral ecology, phylogenetic history, sensory physiology, and reproductive physiology has raised many questions that were not even considered when I began this study. However, these are the details needed to attempt to understand how sexual selection influences an animal's communication system. For some aspects of this study, not enough comparative data are available to allow us to evaluate whether túngara frogs are typical or atypical. And generalizations cannot, or at least should not, be made from one study. Therefore, this study reveals how sexual selection might influence the evolution of behavior of one species. Clearly, we need more detailed studies of more species before we really understand the full potency, or perhaps the impotency, of sexual selection. I hope that my study of túngara frogs has contributed to the available data. I also hope that this study has indicated the necessity of a broad, integrative approach to problems of behavioral evolution. We might be interested in how one behavior functions or evolves, but we must remember that this behavior is part of an entire organism—it is regulated by a nervous system, a physiology, and a morphology; it has functions but it also has consequences, and it has a history. An integrative approach might not be convenient, but if we ever hope to find the right answer to a question as complicated as that of behavioral evolution, we have to ask all the right questions.

Appendix

Methods

This study of the reproductive behavior of *Physalaemus pustulosus* includes investigations of the species' mating system, communication system, and reproductive physiology. Consequently, the techniques employed are routinely used in the fields of behavioral ecology, ethology, animal communication, sensory physiology, and physiological ecology. A knowledge of the precise details of the techniques is not necessary for an understanding of the results and discussions of this study. However, in this appendix I outline the details for those interested in the specific methodologies. Because I presented a brief description of the methods employed prior to reporting the results of various aspects of the research, there is some overlap with previous chapters.

Measuring Male Mating Success

To quantify male mating success, the individual identity of males must be known, and ideally the observer should be able to identify the individuals without disturbing them. Toe clipping is a technique that has been used successfully for marking individuals in a variety of amphibians and reptiles. By removing a portion of a digit, for a combination of digits unique to each frog, individuals can be identified upon recapture. All frogs captured in this study were given a unique toe clip. Although toe regeneration occurs in some (mostly primitive) species of frogs (Ferner 1979), it did not appear to occur in *P. pustulosus* during the course of this study.

Monitoring behaviors of known individuals without having to recapture them for identification was necessary for this study. Emlen (1968a) proposed marking frogs with numbered waistbands for behavioral observations. This and a variety of other techniques, such as using adhesives to stick tags to the frogs (e.g., Ferner 1979), were not successful for *P. pustulosus*. Finally, I sewed small (0.5 by 0.5 centimeters) numbered pieces of surveyor's tagging to the frogs' backs. This did not appear to affect the males' behavior, as they often resumed calling shortly after they were marked and returned to the pool.

Each night I captured all frogs that did not have tags. They were given a unique toe clip, their size was measured (snout-to-vent length), and they were given a tag. Males were measured to the nearest 0.5 millimeters, but for most analyses size was rounded to the nearest millimeter. Some of the frogs captured had previously been marked, but they had lost tags. I recorded the toe-clip pattern of these individuals, and they were remeasured and given a new tag. This procedure also provided some information on growth rates of individuals.

All the males in the study were marked, but not the females. Females were often seen only during amplexus, and thus could not be marked without disturbing the mated pair. In some species, a mated pair that is separated will immediately clasp again when given the opportunity. This was not true for *P. pustulosus*. Therefore, females were marked only if they were captured before mating.

Males began calling at dusk and chorusing usually subsided around midnight. From 1900 to 2400 hours I surveyed the pool, recorded the behavior and location of all males, and collected any males without tags. Males often remained in the calling position between calling bouts; those that were inflated during the censuses probably had been calling recently. Males in the deflated position probably had not been calling immediately prior to censusing, but may have been calling at other times. A male was noted as either inflated (calling), deflated (not calling), clasping a female (mated), or nesting. This procedure probably underestimated the amount of calling, since deflated males might have called between censuses, but the census technique provides an unbiased estimate of the relative amount of calling by different males. Mated males were almost always observed nesting later that night, and nesting was used as the estimate of male reproductive success.

Although the behavior of males was systematically recorded each hour from 1900 to 2400 hours, each time a mated or nesting pair was observed it was recorded immediately. Most nesting took place after 2400 hours. From 2400 hours until all the breeding activity had ceased, usually before 0400 hours and often by 0230 hours, the pool was monitored continually and all males involved in nesting were recorded. Occasionally, the male of a nested pair did not have a tag. Usually this was a previously marked male who lost his tag. In these cases, I placed a small screen cage around the nesting pair. After nesting the pair was retrieved from the cage and marked.

A Pearson product-moment correlation (Steele and Torrie 1960) or a Spearman rank correlation (Siegel 1956) was used to test the hypothesis that there was a correlation between male traits and behaviors, and male mating success. A chi-square analysis (Sokal and Rohlf 1969) was used to test the hypothesis that on any given night the size of mated males was a random sample of the male population at the breeding site on that night.

Acoustic Recording and Analysis

As pointed out in chapter 1, an important advantage of studying anuran reproductive behavior is that the primary sexual display is the advertisement call; using bioacoustic techniques, this sexual display can easily be recorded and analyzed.

The calls of male *P. pustulosus* were recorded with a Nagra IV-D tape recorder and either a Sennheiser MKH 104 or 815 T microphone. This tape recorder with either microphone has a frequency response curve that is fairly flat (within a few decibels) from less than 100 Hz up to 15 to 20 kHz. The microphone was usually placed about 50 centimeters from the frog being recorded, and care was taken not to overload the microphone input level of the tape recorder.

The calls of males were analyzed with a Kay Sonograph, either model 6061B or 7029A, using narrow band frequency filter and a frequency range of 40 to 4000 Hz. The fundamental frequency of the chuck was determined by measuring the distance in Hz between the three lowest chuck harmonics and dividing by three. Additional call analyses were conducted with a Nicolet 444A frequency spectrum analyzer and a Tektronix 5111 storage oscilloscope. The intensity of calls in the field was measured with a General Radio model 1982 Precision Sound Level Meter at 50 centimeters from the calling frog.

Due to the different functions of the whine and the chuck, I was interested in whether these two call components exhibited different attenuation rates. A whine plus chuck was broadcast from a Nagra-Kudelski speaker with a Nagra IV-D tape recorder at an intensity of 90 dB SPL at 0.5 meters from the speaker. The calls were recorded with a second Nagra tape recorder and a Sennheiser MKH 104 microphone at distances of 0.5, 1, 5, 10, and 20 meters from the speaker. The microphone output level was never overloaded, but it was readjusted at each distance in order to maximize the distance over which I was able to record the call. The absolute intensity of the recorded call was not important for this analysis because I was comparing the relative differences in amplitude between the whine and the chuck. The null hypothesis was that the relative difference between the amplitudes does not change with distance. The peak relative amplitude, in volts, of the whine and chuck was determined with a Tektronix 5111 storage oscilloscope. Differential attenuation of the call components was determined by computing the difference between the maximum amplitude of the whine and chuck at each distance. This procedure was carried out in three different habitats.

When present at the breeding site, males spend a large part of the time calling. For a number of reasons, it was necessary to know how many calls a male produced in a single night at the breeding site. Five males, each on different nights, were placed in plastic buckets (approximately 15 centimeters in diameter) with screened tops and a few centimeters of water. The buckets were placed near the *P. pustulosus* chorus behind Kodak House on

Barro Colorado Island. A Sennheiser MKH 104 microphone from a Nagra IV-D tape recorder was placed on top of the screen, directly above the frog. The calls produced by the male from 1900 to 0200 hours were recorded, and the output of the tape recorder was monitored with a General Radio type 1521 graphic chart sound level recorder. A timer was used to activate the system at 1900 hours and to turn it off at 0200 hours. The printout from the chart level recorder allowed us to count the total number of calls produced by the frog during the night.

Preparation of the Acoustic Stimuli

Analysis of the advertisement call of various males revealed a significant negative correlation between male size and the fundamental frequency of the chuck. Although other call characteristics varied among males, only the frequency of the chuck varied predictably with size. To test the hypothesis that females were attracted preferentially to calls with lower frequency chucks, it was necessary to present females with chucks of different frequencies but with the same whine. This was only possible using synthesized calls. Calls were synthesized in which the whine components were identical, but the fundamental frequency of the chuck was varied. The whine had an onset of about 1000 Hz, and in 400 milliseconds the frequency swept linearly to about 400 Hz. The chuck temporally overlapped the whine. The calls had only one chuck, and the chuck had energy in twelve harmonics of a fundamental frequency that varied from 200 Hz to 260 Hz.

Synthesized pure tones were used in some of the phonotaxis experiments with females. These tones were 400 milliseconds in duration. In the other playback experiments with both males and females, natural calls were used. Typically, a continuous tape loop of a natural call was produced. A reversed whine was produced by playing a natural whine backwards on the tape recorder. This resulted in the same frequency-energy profile being produced as in a natural call, but in the reversed direction. A chuck by itself does not occur in nature and was produced by dissecting a chuck from a natural call. Likewise, a chuck followed by a whine is not a natural call and was produced by dissecting a chuck from a whine plus chuck and splicing it immediately in front of a whine.

Preferential Phonotaxis of Females

An important technique in the study of anuran behavior, first used by Littlejohn and Michaud (1959), is the phonotaxis experiments to test female preference among advertisement calls. The design of these experiments was straightforward. A female was placed between two speakers with each speaker broadcasting a different stimulus. Earlier studies usually presented females with a heterospecific and a conspecific advertisement call. After a sufficient number of trails, whether one of the stimuli preferentially elicited phonotaxis from a female could be determined. Females usually preferred the conspecific call. These studies demonstrated the importance of the anuran advertisement call as a primary species-isolating mechanism. I employed a similar technique to investigate the preference of females among conspecific calls. The test arena was octagonal in shape and had a burlap wall about 1 meter high. Two extension speakers, each driven by a Nagra IV-D tape recorder, were placed outside the burlap wall, on opposite sides of the arena, 150 centimeters apart, and faced towards the center of the arena. A 40-watt red light bulb with a rheostat control was located 1.5 meters directly above the center of the arena and was used to maintain a light level in which the female was barely visible to the observer.

The stimuli were broadcast alternatively from the speakers at a rate of about 1 stimulus every 2.3 seconds. Therefore, the frog received a stimulus once every 1.15 seconds. The intensity of the stimuli were equalized at the center of the arena, 75 centimenters from each speaker, at 70 dB SPL using a Realistik sound level meter with fast response. This meter was later calibrated with a General Radio model 1982 sound pressure level meter using túngara frog calls and pure tones for the calibration. To check for acoustic anomalies within the arena, I measured the sound intensity at different distances from each speaker, and in each case the sound intensity was within 1 or 2 dB SPL of that predicted by the inverse square law. Thus the difference in sound pressure levels within the arena was primarily due to the spherical spreading of sound. The test arena was located in a large screened room, and the temperature was the same as that outdoors, usually within 1.5° of 25° C.

Testing was initiated by placing a female in the center of the arena in a small cage, either inside a wire cylinder or under a small funnel. After the female was introduced into the arena, stimulus presentation began. The stimuli were broadcast for 30 seconds, the cage removed from the female, and her response recorded. A "speaker response" was noted if she contacted the burlap wall within 12 centimeters of one of the speakers. A "no response" occurred if the female did not contact the wall during the 3 to 5 minutes the stimuli were played or if she contacted the wall outside of the 12-centimeter area around each speaker.

The stimuli were switched between speakers on successive trials and between tape recorders on successive nights. Females were tested only once with each stimulus pair. Statistical significance for preferential phonotaxis was tested with a one-tailed exact binomial probability test (Sokal and Rohlf 1969). The one-tailed test was used because there was an a priori prediction, based on the measures of male mating success, as to the direction of the female preference, that is, that females should prefer lower frequency calls (Sokol and Rohlf 1969).

Evoked Vocal Responses of Males

I also compared the ability of different stimuli to elicit vocal responses from male *P. pustulosus*. Stimuli were tested in pairs, and the responses of males to each stimulus of the pair were compared with a Wilcoxon matched-pairs signed-ranks test (Siegel 1956). The male responses quantified were: (1) total number of calls, (2) total number of chucks, (3) time to first call, and (4) time to first chuck.

A male was placed in a small dish of water with a screened top in the center of the test arena described above. Stimuli were broadcast with a Nagra IV-D tape recorder and a small extension speaker outside the burlap wall and 75 centimeters from the male. The intensity of the stimulus at the male was 75 dB SPL. A male was presented with one type of stimulus for 30 seconds at a rate of 1 stimulus every 2.3 seconds. The male's response was recorded simultaneously with a Nagra IV-D tape recorder and a Sennheiser MKH 815 T microphone. Several minutes after the stimulus presentation was terminated the male was presented with the second stimulus of the stimulus pair, and his response was recorded in the same manner. Each male was tested only once with each stimulus pair, and the stimulus of each pair that was presented first was alternated among males.

Male Contribution to Female Reproductive Success

Female behavior clearly influenced male reproductive success. Some researchers have also suggested that the size of a male frog might influence the reproductive success of the female with which he mates (e.g., Davies and Halliday 1977; Howard 1978a; see also chap. 1). Therefore, investigating nesting behavior of *P. pustulosus* was necessary. Túngara frogs deposit their eggs in foam nests (Heyer and Rand 1977; chap. 3). Because pairs often construct nests communally, determining the fertilization or developmental success of clutches from known pairs in the field is not possible. To determine the possible influence of male size on female reproductive output, I measured the size of individual males and females and then placed them in small plastic buckets in the laboratory, one pair in each bucket. Most pairs (> 90%) mated, constructed nests, and deposited eggs. Eggs usually hatched in three days. After four or five days the number of tadpoles from each nest was recorded. The foam nest was then broken down with a weak chlorine solution (approximately 5 water : 1 chlorine bleach), and the number of undeveloped eggs was counted. Nests from the field were treated in a similar manner to compare the proportion of undeveloped eggs per nest in the field and in the laboratory.

Data were analyzed by nonparametric statistics (Siegel 1956) except in the regressions where the parametric assumptions were met. Because the number of undeveloped eggs per nest was distributed in a Poisson-like manner, these data were transformed by square root (Steele and Torrie 1960). The multiple regression analysis was computed with the Statistical Analysis System (SAS Institute, Raleigh, North Carolina). The importance of individual variables in the regression equations was determined by comparing the sum of squares of each variable to the residual mean sum of squares with an F test (Steele and Torrie 1960).

Energetics of Reproductive Behavior

Although the potential benefits of sexual advertisement are obvious, few data are available to evaluate the potential costs of this behavior. One of the important costs to consider is energy expenditure. We estimated the aerobic (as measured by oxygen consumption) and the anaerobic (as measured by lactate buildup) energy expenditures of various reproductive behaviors of males and females. For comparative purposes, we also determined energy costs during resting and forced activity. The techniques for determining energy expenditure can be complex, but the principle involved is straightforward. The amount of oxygen consumed can be related directly to the number of calories produced. If the amount of oxygen consumed during a particular behavior is determined, calculating the calories (or joules) produced—which is the aerobic energy cost—is trivial. Similarly, the anaerobic energy cost can be quantified by determining lactate buildup during a behavior, since lactate is the primary metabolic product of anaerobic energy metabolism.

Oxygen consumption was measured in a closed system using an Applied Electrochemistry Oxygen Analyzer. The respirometer chambers were cylindrical glass jars 9 centimeters in diameter and 14 centimeters high, sealed with two-hole rubber stoppers, and equipped with Tygon tubing, three-way valves, and connectors.

All measurements were made in the dark at ambient temperatures (25° C to 27° C). Daytime resting measurements were made between 0800 and 1300 hours. All other measurements were made between 1930 and 0500 hours. Nighttime resting measurements were made in a laboratory, where the experimental animals could not hear the calling of the other frogs.

Each measurement of rate of oxygen consumption ($\dot{V}O_2$) was paired with a simultaneous measurement of the oxygen concentration in the control respirometer in which there was no frog. An hour or more before each run, a measured volume of water at ambient pressure was added to a pair of jars, and a frog was introduced into one of them. One to several hours later the jars were sealed, incurrent and excurrent valves were closed, and a timer started. The periods of measurement were approximately 0.5 hours for calling and nesting frogs, 1 hour for noncalling frogs at night, and 4 hours for resting frogs in the daytime.

The fractional concentration of oxygen (FO_2) in the air from each jar at the end of the period of measurement was determined by introducing a regulated flow of water (in equilibrium with room air) from an overhead reservoir. This water displaced the air from the respirometer jars at a constant rate (usually 25 milliliters/minute) and forced the air sample through a scrubber containing Drierite and Ascarite to remove water vapor and CO_2, through the oxygen sensor, and through a flow meter. Rate of oxygen consumption was calculated from the following formula and corrected to standard temperature and pressure:

$$\dot{V}O_2 = \frac{V(FIO_2 - FEO_2)}{(1 - FEO_2)\ t}$$

where V is the volume of dry air in the respirometer chamber, FIO_2 is the initial fractional concentration of oxygen in the air as determined from the sample in the blank respirometer, FEO_2 is the fractional concentration of oxygen in the air at the end of the run, and t is the duration of the period over which the measurements were made.

The procedure described above was modified to obtain metabolic measurements from calling frogs. Twenty-four hours or more before measurements, several frogs were taken from the chorus behind Kodak House and put into the respirometer jars. The jars were filled with water to a depth of 2.5 centimeters, covered with a wire mesh screen, and placed on a visually shielded bench approximately 3 meters from the pond. Thus the frogs could hear the calls of the chorus from which they had been removed, and they were subject to the same environmental temperature and photoperiod as the animals in and around the frog pond. At dusk three or four jars with the frogs in them were sealed and attached in series with a control jar, and air was drawn through the system at a rate of 650 milliliters/minute. At this flow rate no measurable reduction in the fractional concentration of oxygen developed in the respirometer jars, and the fractional concentration of oxygen in the control jar was indistinguishable from that in the experimental respirometers.

The respirometer jars containing the calling frogs were located from 0.2 to 1.0 meters from a speaker continuously broadcasting tape-recorded advertisement calls of túngara frogs. The recorded call, a whine plus chuck, was repeated at various intensities at a rate of 1 call every 2.3 seconds, using a Nagra IV-D tape recorder and a Nagra-Kudelski speaker. As soon as one of the frogs inflated, floated in the water, and began to call, its respirometer and the blank respirometer jar were connected in series with a pump and a flow meter. A Sennheiser MKH 815 T or MKH 104 microphone was placed adjacent to the respirometer jar containing the frog, and the vocalizations of the calling male in the jar were recorded on a Nagra IV-D tape recorder. Because the microphone was immediately adjacent to the respirometer, the calls produced by an experimental frog were clearly distinguishable from those of the playback tape and the background chorus. After a few minutes of auditory stimulation from the speaker and the nearby frog chorus, the frog in the respirometer usually resumed calling. When it did so, the incurrent and excurrent valves to the paired experimental respirometer jars were closed, which converted them into two separate systems. The time was noted, and the tape recorder was turned on. After 30 minutes, the tape recorder was turned off, the respirometer jars were taken to an adjacent laboratory, and the fractional concentration of oxygen was determined as previously described. The tapes were later analyzed to determine both the number and the type of calls produced during the experiment. Some frogs never resumed calling after their respirometers were sealed, remaining inactive throughout the experimental periods (fig. A.1).

We used a modified form of the above method to measure oxygen consumption during nesting. Between 1730 and 1830 hours a female and a male, which had been captured at a breeding pond near Gamboa the previous night,

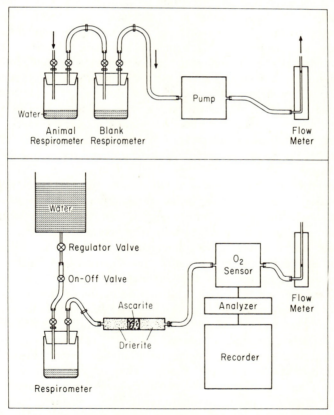

Figure A.1 A diagram of the apparatus used to measure the rate of oxygen consumption in *Physalaemus pustulosus* during various behavioral activities (from Bucher, Ryan, and Bartholomew 1982).

were placed in an upstream jar. After several minutes the frogs usually clasped, and thereafter they were watched continuously. About half of the amplexed pairs constructed nests. As soon as nest building started, the valves to the respirometer jar and the blank jar were closed. When nest building was completed the frogs were separated, and some of them were used for lactic acid determinations. Instead of using a reservoir to force the air from the chambers, as we did in the calling-frog experiments, samples of air were withdrawn from the experimental and control respirometers with a 60-cubic-centimeter syringe equipped with a three-way valve. The syringes were placed in a Razel syringe pump, and the air they contained was expelled into the oxygen analyzer at a rate of 20 milliliters/minute. The remainder of the oxygen analysis and the computations were as previously described.

The rate of oxygen consumption during forced activity was determined in a similar manner, but the initial gas sample was room air taken into the syringe at the same time and place that the experimental respirometer was filled with air. The respirometer was a plexiglas syringe (internal diameter 5 centimeters) containing several drops of water to keep the animal saturated. The frog was placed in the syringe and the volume adjusted to 100 milliliters. A stopwatch was started, and the syringe was rotated by hand, forcing the frog either to move or right itself. At the end of the period of forced activity, the time was noted, and a sample of air in the syringe respirometer was transferred through a pair of valves into a 60 cubic centimeter syringe. Fractional concentrations of oxygen in the blank and experimental syringes were determined as described above. At the completion of nest building and at the end of the period of forced activity, the frogs were killed for lactate determination.

The contributions of anaerobic metabolism to calling, nest building, and forced activity were assayed by lactic acid determinations of whole-animal homogenates. Immediately after each experiment, the frog was decapitated and homogenized in a Waring blender in a volume of chilled 0.6 molar perchloric acid. A sample of each homogenate was kept chilled in a sealed vial, and the samples were returned to the University of California, Los Angeles for centrifugation. The supernatant was processed using a Calbiochem-Behring lactate kit, and the lactate concentration was determined at 300 nanometers with a Beckman model 35 spectrophotometer.

Male túngara frogs captured in Gamboa were used to determine the amount of lactate accumulation during calling. These males were placed in individual screen-covered jars 9 centimeters in diameter and 14 centimeters tall containing water to a depth of 3 or 4 centimeters. The jars were then placed on a bench approximately 3 meters from the Kodak House frog pond where the experimental animals could hear the nocturnal frog chorus. After they had been calling for 3 to 4 hours, they were killed for lactate determination. A Sennheiser MKH 104 microphone and a Nagra IV-D tape recorder were used to record the calls produced by the experimental males.

During the experiments the output of the tape recorder was monitored with the General Radio graphic chart level recorder. The printout of the chart recorder allowed us to count the number of calls produced by the experimental males. Although the printout did not reliably distinguish the number of chucks per call, at the time of these experiments we already were aware that the aerobic energy expended per call was not influenced by the number of chucks. Therefore, call complexity (i.e., number of chucks) was not quantified in this portion of the study.

The energetic contribution of females to the clutch also was determined. Nests produced by four of the experimental pairs and one nest collected in the field were weighed and dried in a forced-air oven at approximately 50° C for at least 24 hours immediately after they were collected. At Cornell University, the caloric content of the dried eggs was determined with a Parr Instrument Company oxygen bomb calorimeter.

A Student's t test was used to compare energetic costs of different behavioral states and differences between the two sexes. Regression equations were fitted by the method of least squares (Steele and Torrie 1960).

Preferential Response by Frog-eating Bats to Túngara Frog Calls

The discovery that the bat *Trachops cirrhosus* (Phyllostomatidae) eats frogs and uses the frogs' vocalizations as locational cues allowed a direct test of the hypothesis that sexual advertisement incurs a predation cost. Bats used in the experiments were mist-netted along streams on Barro Colorado Island. Preference of the bats to various pairs of túngara frog calls was tested in the field and in a large outdoor flight cage. The flight cage measured 4 meters on each side and 2.1 meters high. It had a tin roof, and the walls were screen. The bat perched in one corner and the observer sat in the opposite corner. One speaker was located in each corner adjacent to the bat; the distance from each speaker to the bat was 4 meters.

An individual bat was presented with a pair of *P. pustulosus* calls which were broadcast simultaneously with Stellavox tape recorders and small extension speakers. The intensity of the stimulus calls was 75 dB SPL at 1 meter from the speaker, and the call repetition rate was 1 call every 1.6 seconds. A response was recorded if a bat flew within 1 meter of the speaker during the 60 seconds of stimulus presentation. To avoid habituation the calls were discontinued immediately after the bat left the perch. No rewards were given, and the stimuli were switched between speakers after each trial. Bats were usually tested until they did not respond in several consecutive trials; no bats were tested in more than eleven trials in a given night.

The response of the bats to *P. pustulosus* calls in the field was also determined. Two Pearlcorder model D 120 microcassette recorders broadcast frog calls at sites of known *T. cirrhosus* activity. Calls were broadcast simultaneously at an intensity of 74 dB SPL at 1 meter from the speaker. The speakers were placed 4 meters apart from each other, and were observed simultaneously with a Javelin model 221 night vision scope from a distance of about 12 meters. Responses were recorded when a *T. cirrhosus*, which we can identify on the wing, flew within 1 meter of a speaker.

The number of responses by each bat in the flight cage, or the total number of passes on each night, were not equal. Therefore, results for each set of experiments were analyzed by computing the exact binomial probability of a random choice by each bat (flight cage) or on each night (field tests). The null hypothesis of no preference was then tested by comparing $-2 \Sigma \ln P$ to a chi-square distribution (Sokal and Rohlf 1969).

The Response of Túngara Frogs to the Frog-eating Bat

We determined the effect that the presence of the frog-eating bat had on the vocal behavior of male túngara frogs. Calls of a chorus of túngara frogs were recorded continuously at Weir Pond from 1830 to 2000 hours on each of six nights with a Nagra IV-D tape recorder and a Sennheiser MKH 104 microphone. Observations of *T. cirrhosus* were made with a Javelin night vision scope. We recorded the time of visits by *T. cirrhosus* to the nearest second on a Pearlcorder microcassette tape recorder. The temporal distribution of frog calls was later quantified using a graphic chart level recorder, and times of bat visits to the pond were transcribed from the Pearlcorder tape to the printout of calls from the chart recorder. We censused the number of calling túngara frogs at the beginning and end of each night of observation.

To determine the effect of bats on the calling frogs, we counted the number of calls in the 15-second period immediately before and after the arrival of each bat. The maximum number of calls that could be discriminated on the chart recorder was 65. The null hypothesis of no effect predicted that the periods before and after a bat arrival should contain the same number of calls. The data were recorded on selected nights under one of the following light conditions: (1) full moon, clear, (2) full moon, cloudy, (3) no moon, clear, and (4) no moon, cloudy. Incident light levels for each night were measured with a Gossen Lunasix electronic exposure meter. We used a Wilcoxon matched-pairs signed-ranks test to compare the number of calls before and after each bat visit (Siegel 1956).

We used a model of *T. cirrhosus* to examine in detail the behavioral response of frogs to the presence of bats. These experiments were conducted at Kodak Pond under varied light conditions. Models of *T. cirrhosus* and a small insectivorous bat, *Micronycteris megalotis*, were made by tracing an outline of an individual of each species with its wings spread. The outline was then cut out on heavy cardboard, stained to approximately the bat's color, weighted below, and rigged above with a small roller. The models were passed diagonally across the pond on a taut, monofilament line. Speed was adjusted to approximately normal *T. cirrhosus* flight speed by adjusting the incline of the line. A wingless control model, mechanically identical to the *T. cirrhosus* model, was used to test for the possibility that frogs responded to sounds generated by the model. In these tests the models were passed over the pond beginning at a maximum height of 1.0 meter and ending at 0.2 meters. One end of the line was always slightly lower to provide a slope for the model to roll down. The observer was 4 meters from the pond and out of the frogs' view.

Because the frogs occasionally ceased calling in the absence of any apparent threat or disturbance, we recorded data during a 15-minute control period prior to each experimental trial. The time from cessation of calling by the frogs (i.e., a shutdown) to resumption of full chorusing was noted during each control period. We ran each test after a full chorus had been in session for at least 30 seconds following the 15-minute control period. If the frogs ceased calling within the time required for the model to traverse the pond, we assumed that the cessation of calling was in response to the model. We compared the duration of chorus shutdowns during the control period to the shutdowns in response to the bat model with a Wilcoxon signed-ranks test (Siegel 1956).

At least five experimental trials and an equal number of controls were conducted during each night of experimentation. We performed the following tests between 1930 and 2400 hours: (1) *T. cirrhosus* model was presented at intermediate light intensities (incident light levels approximately 0.17 and 0.35 lux), and in "total darkness" (incident light level below 0.17 lux, no moon, heavy overcast); (2) wingless model was presented at intermediate light levels; and (3) *T. cirrhosus* model and a *M. megalotis* model were presented at incident light intensities of 0.50 to 1.40 lux. In the third comparison, we ran the two models separately for each trial and alternated models between trials. Thus the *T. cirrhosus* model was released first one time and second the next. The second model was run as soon as the frogs had restarted a full chorus for 30 seconds, but 15-minute control periods were run between pairs of trials. Statistical comparisons of the various treatments were computed as described above.

We also tested the responses of individual frogs to varied heights (1.00 to 0.13 meters) of the *T. cirrhosus* model. As the model passed along the downward sloping monofilament line, height decreased from one side of the pond to the other. The model was run along both the east and west sides of the pond, 9 times headed north and 9 times headed south. Model runs were conducted at approximately 10-minute intervals from 2000 to 2230 hours.

Both east and west sides of the pond were marked every 30 centimeters. Prior to each run, we noted the position and posture of each calling male on a diagram of the pond. The model was passed over the calling frogs as soon after a 10-minute interval ended when three consecutive counts under a dim headlight consistently showed the same numbers and positions of calling frogs in each marked section.

Immediately after the model passed over the frogs (the model was released by remote control), a dim headlight was used to note changes in body posture of each frog. Frog responses to the model were characterized as follows: (1) continued calling, (2) stopped calling but remained inflated, (3) stopped calling and deflated, (4) deflated and lowered body until only the snout was protruding from the surface, and (5) dived. After behavioral observations and postural changes were noted, the model was slowly passed back along the monofilament line, and a meter stick was used to measure model height above each frog. A Spearman rank correlation test was used to test for an association between the frog's response and the height of the model (Siegel 1956).

References

Alexander, R. D. 1974. The evolution of social behaviour. *Ann. Rev. Ecol. Syst.* 6:325–83.
———. 1975. Natural selection and specialized chorusing behavior in acoustical insects. In *Insects, society, and science*. New York: Academic Press.
Altig, R. 1979. *Toads are nice people*. Columbia and Eldon, Mo.: Gates House/MANCO.
Anderrson, M. 1982a. Female choice selects for extreme tail length in a widowbird. *Nature* 299:818–20.
———. 1982b. Sexual selection, natural selection, and quality advertisement. *Biol. J. Linn. Soc.* 17:375–93.
Andrews, R. M., and Rand, A. S. 1974. Reproductive effort in anoline lizards. *Ecology* 55:1317–27.
Arak, A. 1983a. Mating behaviour of anuran amphibians: The roles of male-male competition and female choice. In *Mate choice*, ed. P. Bateson. Cambridge: Cambridge Univ. Press.
———. 1983b. Sexual selection by male-male competition in natterjack toad choruses. *Nature* 306:261–62.
Arnold, S. J. 1981. Behavioral variation in natural populations, 1: Phenotypic, genetic, and environmental correlations between chemoreceptive responses to prey in the garter snake, *Thamnophis elegans*. *Evolution* 35:489–509.
———. 1983. Sexual selection: The interface of theory and empiricism. In *Mate choice*, ed. P. Bateson. Cambridge: Cambridge Univ. Press.
Axtell, R. W. 1958. Female reaction to the male call in two anurans (Amphibia). *Southwest. Nat.* 3:70–76.
Baker, R. R., and Parker, G. A. 1979. The evolution of bird colouration. *Phil. Trans. Royal Soc. London* 287:63–130.
Barrio, A. 1953. Sistematics, morphologia, y reproducion de *Physalaemus henseli* (Peters), y *Pseudopalidicola faliceps* (Anura, Leptodactylidae). *Physis* 20:379–89.
———. 1964a. Relaciones morfologicas, eto-ecologia, y zoogeographica

entre *"Physalaemus henseli"* (Peters), y *"P. fernandazea"* (Muller) (Anura, Leptodactylidae). *Acta Zoologica Lilloana* 20: 285–305.

———. 1964b. Importancia, significacion, y analisis del canto de batracios anuros. In *Seperado de la publicacion comemorativa del Cincuentavo del Museo Provincial de Cincias Naturales "Florentino Ameghino"*. Sante Fe, Argentina.

———. 1965. El genero *Physalaemus* (Anura, Leptodactylidae) en la Argentina. *Physis* 25:421–48.

———. 1966. Divergencia acustica entre el canto nupcial de *Leptodactylus ocellatus* (Linne) y *L. chaguensis* Cei (Anura, Leptodactylus). *Physis* 26:275–77.

———. 1967. Notas complementarias sobre el genero *Physalaemus* Fitzingeri (Anura, Leptodactylidae). *Physis* 27:5–8.

Bartholomew, G. A. 1970. A model for the evolution of pinniped polygyny. *Evolution* 24:546–59.

Bateman, A. J. 1948. Intrasexual selection in *Drosophila*. *Heredity* 2:349–68.

Bateson, P., ed. 1983. *Mate choice*. Cambridge: Cambridge Univ. Press

Bennett, A. F., and Licht, P. 1974. Anaerobic metabolism during activity in amphibians. *Comp. Biochem. Physiol.* 48A:319–27.

Beranek, L. L. 1954. *Acoustics*. New York: McGraw-Hill.

Berven, K. A. 1981. Mate choice in the wood frog, *Rana sylvatica*. *Evolution* 35:707–22.

———. 1982. The genetic basis of altitudinal variation in the wood frog *Rana sylvatica*, 1: An experimental analysis of life history traits. *Evolution* 36:962–83.

Blair, W. F. 1955. Size differences as a possible isolating mechanism in *Microhyla*. *Amer. Natur.* 89:297–301.

———. 1956. Call differences as an isolating mechanism in southwestern toads. *Texas J. Science* 8:87–106.

———. 1958a. Response of a green treefrog *(Hyla cinera)* to the call of the male. *Copeia* 1958:333–34.

———. 1958b. Mating call in the speciation of anuran amphibians. *Amer. Natur.* 92:27–51.

———. 1964. Isolating mechanisms and interspecies interactions in anuran amphibians. *Quart. Rev. Biol.* 39:334–44.

———, ed. 1972. *Evolution in the genus Bufo*. Austin: Univ. Texas Press.

Blair, W. F., and Littlejohn, M. J. 1959. Stage of speciation in two allopatric populations of chorus frogs *Pseudacris*. *Evolution* 14:82–87.

Bogert, C. M. 1960. The influence of sound on the behavior of amphibians and reptiles. In *Animal sounds and communication*, ed. W. E. Langyon and W. N. Tavolga. A.I.B.S. Pub. 7:137–320.

Borgia, G. 1979. Sexual selection and the evolution of mating strategies. In *Sexual selection and reproductive competition in insects*, ed. M. S. Blum and N. H. Blum. New York: Academic Press.

Bowman, R. I. 1979. Adaptive morphology of song dialects in Darwin's finches. *J. Ornithol.* 120:353–89.

Brackenbury, J. H. 1977. Physiological energetics of cock-crow. *Nature* 270:433–35.
———. 1982. The structural basis of voice production and its relationship to sound characteristics. In *Acoustical communication in birds, 1: Production, perception, and design features,* ed. D. E. Kroodsma and E. H. Miller. New York: Academic Press.
Bradbury, J. W. 1977. Lek mating behavior in the hammer-headed bat. *Z. Tierpsychol.* 45:225–55.
———. 1981. The evolution of leks. In *Natural selection and social behavior: Recent research and new theory,* ed. R. D. Alexander and D. W. Tinkle. New York: Chiron Press.
Bradbury, J. W. and Gibson, R. M. 1983. Leks and mate choice. In *Mate choice,* ed. P. Bateson. Cambridge: Cambridge Univ. Press.
Bradbury, J. W., and Verhencamp, S. L. 1977. Social organization and foraging in emballonurid bats: Mating systems, 3. *Behav. Ecol. Sociobiol.* 2:1–17.
Brattstrom, B. H., and Yarnell, R. M. 1968. Aggressive behavior in two species of leptodactylid frogs. *Herpetologica* 24:222–28.
Bray, O. E.; Kennedy, J. J.; and Guarino, J. L. 1975. Fertility of eggs produced on territories of vasectomized red-winged blackbirds. *Wilson Bull.* 87:187–95.
Brenowitz, E. A. 1982. Long-range communication of species identity by song in the red-winged blackbird. *Behav. Ecol. Sociobiol.* 10:29–38.
Brown, C. H. 1982. Ventroloquial and locatable vocalizations in birds. *Z. Tierpsychol.* 59:338–50.
Brown, J. L. 1964. The evolution of diversity in avian territorial systems. *Wilson Bull.* 76:160–69.
———. 1982. The adaptationist program. *Science* 217:884–85.
Bucher, T. L.; Ryan, M. J.; and Bartholomew, G. A. 1982. Oxygen consumption during resting, calling, and nest building in the frog, *Physalaemus pustulosus*. *Physiol. Zool.* 55:10–22.
Buechner, H. K., and Roth, H. D. 1974. The lek system in Uganda kob antelope. *Amer. Zool.* 14:145–62.
Cade, W. H. 1975. Acoustically orienting parasites: Fly phonotaxis to cricket song. *Science* 190:1312–13.
———. 1981. Alternative male strategies: Genetic differences in crickets. *Science* 212: 563–64.
Campbell, B., ed. 1972. *Sexual selection and the descent of man.* Chicago: Aldine Press.
Cannatella, D. C., and Duellman, W. E. 1984. Leptodactylid frogs of the *Physalaemus pustulosus* group. *Copeia* 1984:902–21.
Capranica, R. R. 1965. *The evoked vocal response of the bullfrog.* Res. Monogr. No. 33. Cambridge: M.I.T. Press.
———. 1966. Vocal response of the bullfrog to natural and synthetic mating calls. *J. Acoustic. Soc. Amer.* 40:1131–39.
———. 1976a. The auditory system. In *Physiology of the amphibia* 3, ed. B. Lofts. New York: Academic Press.

---. 1976b. Morphology and physiology of the auditory system. In *Frog neurobiology,* ed. R. Llinas and W. Precht. Berlin: Springer-Verlag.

---. 1977. Auditory processing of vocal signals in anurans. In *Reproductive biology of amphibians,* ed. D. H. Taylor and S. I. Guttman. New York: Plenum Pub.

Capranica, R. R.; Frischkopf, L. S.; and Nevo, E. 1973. Encoding of geographical dialects in the auditory system of the cricket frog. *Science* 182:1272–75.

Capranica, R. R., and Moffat, A. J. M. 1975. Selectivity of the peripheral auditory system of spadefoot toads (*Scaphiopus couchi*) for sounds of biological significance. *J. Comp. Physiol.* 100:231–49.

---. 1980. Nonlinear properties of the peripheral auditory system. In *Proceedings in life sciences: Comparative studies of hearing in vertebrates,* ed. A. N. Popper and R. R. Fay. New York: Springer-Verlag.

---. 1983. Neurobehavioral correlates of sound communication in anurans. In *Advances in vertebrate neuroethology,* ed. J. P. Ewert, R. R. Capranica, and D. J. Ingle. New York: Plenum Pub.

Carey, C. 1979 Aerobic and anaerobic energy expenditure during rest and activity in montane *Bufo b. boreas* and *Rana pipiens. Oecologia* 39:213–28.

Cei, J. 1980. *Amphibians of Argentina.* Suppl. Ital. Zool. Soc.

Christian, K. A., and Tracy, C. R. 1983. Thermoregulation and mate selection in Fowler's toad? *Science* 219:518–19.

Cole, L. C. 1954. The population consequences of life history phenomenon. *Quart. Rev. Biol.* 29:103–37.

Collias, N. E. 1960. An ecological and functional classification of animal sounds. In *Animal sounds and communication,* ed. W. E. Lanyon and W. E. Tavolga. Amer. Inst. Biol. Sci. Pub. No. 7, Washington, D.C.

Cordosa, A. J. 1981. Biologia e sobrevivencia de *Physalaemus cuvieri* Fitz: 1826 (Amphibia, Anura), na natureza. *Ciencia e Cultura* 33:1224–28.

Cott, H. B. 1940. *Adaptive colouration in animals.* London: Methuen.

Crump, M. L., and Kaplan, R. H. 1979. Clutch energy partitioning of tropical treefrogs. *Copeia* 1979:626–35.

Darwin, C. 1859. *The origin of species.* Reprint of original edition. New York: Random House.

---. 1871. *The descent of man and selection in relation to sex.* Reprint of original edition. New York: Random House.

---. 1872. *Expressions of the emotions in men and animals.* New York: Appleton.

---. [1876] 1977. Sexual selection in relation to monkeys. *Nature* 15:18–19. Reprinted in *The collected papers of Charles Darwin,* ed. P. H. Barrett. Chicago: Univ. Chicago Press.

---. [1882] 1977. On the modification of a race of Syrian street dogs: With a preliminary notice by Charles Darwin. *Proc. Zool. Soc. London* 25:367–70. Reprinted in *The collected papers of Charles Darwin,* ed. P. H. Barrett. Chicago: Univ. Chicago Press.

Davidson, E. H., and Hough, B. R. 1969. Synchronous oogenesis in *En-*

gystomops pustulosus, a neotropic anuran suitable for laboratory studies: Localization in the embryo of RNA synthesized at the lampbrush stage. *J. Exper. Zool.* 172:25–48.

Davies, N. B. 1978. Ecological questions about territorial behaviour. In *Behavioural ecology: An evolutionary approach*, ed. J. R. Krebs and N. B. Davies. Sunderland, Mass.: Sinauer Assoc.

Davies, N. B., and Halliday, T. R. 1977. Optimal mate selection in the toad. *Nature* 269:56–58.

―――. 1978. Deep croaks and fighting assessment in toads, *Bufo bufo*. *Nature* 274:683–85.

―――. 1979. Competitive mate searching in male common toads, *Bufo bufo*. *Anim. Behav.* 27:1253–67.

Davis, J. W. F., and O'Donald, P. 1976. Sexual selection for a handicap: A critical analysis of Zahavi's model. *J. Theor. Biol.* 57:345–54.

Davison, G. W. H. 1982. Sexual displays of the great Argus pheasant, *Argusianus argus*. *Z. Tierpsychol.* 58:185–202.

Dominey, W. 1983. Sexual selection, additive genetic variance, and the "phenotypic handicap." *J. Theor. Biol.* 101:495–502.

Downhower, J. F., and Brown, L. 1980. Mate preference of female mottled sculpins, *Cottus baridi*. *Anim. Behav.* 28:728–34.

Drewry, G. E.; Heyer, W. R.; and Rand, A. S. 1982. A functional analysis of the complex call of the frog *Physalaemus pustulosus*. *Copeia* 1982:636–45.

Drewry, G. E., and Rand, A. S. 1983. Characteristics of an acoustic community: Puerto Rican frogs of the genus *Eleutherodactylus*. *Copeia* 1983:941–53.

Duellman, W. E. 1958. A monographic study of the colubrid snake genus *Leptodiera*. *Bull. Amer. Mus. Nat. Hist.* 114:1–57.

―――. 1970. *Hylid frogs of Middle America*, 1 and 2. Monogr. of Mus. Nat. Hist. Lawrence: Univ. Kansas.

―――. 1978. The biology of an equatorial herpetofauna. *Univ. Kansas Mus. Nat. Hist. Pub.*, no. 65.

Duellman, W. E., and Pyles, R. A. 1983. Acoustic resource partitioning in anuran communities. *Copeia* 1983:639–49.

Dunbar, R. I. M. 1976. Some aspects of research design and their implications in the observational study of behavior. *Behaviour* 58:78–98.

Ehrman, L. 1972. Genetics and sexual selection. In *Sexual selection and the descent of man*, ed. B. Campbell. Chicago: Aldine.

Embleton, T. F. W.; Piercy, J. E.; and Olsen, N. 1976. Outdoor propagation over ground of finite impedance. *J. Acoustic. Soc. Amer.* 59:267–77.

Emerson, S. B. 1978. Allometry and jumping in frogs: Helping the twain to meet. *Evolution* 32:551–64.

Emlen, S. T. 1968a. A technique for marking anuran amphibians for behavioral studies. *Herpetologica* 24:172–73.

―――. 1968b. Territoriality in the bullfrog, *Rana catesbeiana*. *Copeia* 1968:240–43.

―――. 1976. Lek organization and mating strategies in the bullfrog. *Behav. Ecol. Sociobiol.* 1:283–313.

Emlen, S. T., and Oring, L. W. 1977. Ecology, sexual selection, and the evolution of mating systems. *Science* 197:215–23.

Endler, J. A. 1980. Natural selection on color patterns in *Poecilia reticulata*. *Evolution* 31:78–91.

Fairchild, L. 1981. Mate selection and behavioral thermoregulation in Fowler's toad. *Science* 212:950–51.

Falconer, D. S. 1981. *Introduction to quantitative genetics*. Edinborough: Oliver and Boyd.

Fellers, G. M. 1975. Behavioral interactions in North American treefrogs (Hylidae). *Chesapeake Sci.* 16:218–19.

———. 1979a. Aggression, territoriality, and mating behaviour in North American treefrogs. *Anim. Behav.* 27:107–19.

———. 1979b. Mate selection in gray treefrogs, *Hyla versicolor*. *Copeia* 1979:286–90.

Feng, A. S.; Narins, P. M.; and Capranica, R. R. 1975. Three populations of primary auditory fibers in the bullfrog (*Rana catesbeiana*): Their peripheral origins and frequency sensitivities. *J. Comp. Physiol.* 100:221–29.

Ferner, J. W. 1979. A review of marking techniques for amphibians and reptiles. *Herp. Circular*, no. 9.

Fisher, R. A. 1958. *The genetical theory of natural selection*. 2d rev. ed. New York: Dover Press.

Freeman, H. L. 1967. Geographic distribution of *Engystomops pustulosus* (Anura: Leptodactylidae). M.S. diss., University of Kansas, Lawrence.

Frishkopf, L. S.; Capranica, R. R.; and Goldstein, M. H. 1968. Neural coding in the bullfrog's auditory system: A teleological approach. *Proc. I. E. E. E.* 56:969–80.

Fuzessery, Z. M., and Feng, A. S. 1981. Frequency representation in the dorsal medullary nucleus of the leopard frog, *Rana p. pipiens*. *J. Comp. Physiol.* 143:339–47.

Gallardo, J. M. 1970. Estudio ecologica sobre los anfibios y reptiles del sudoeste de la provincia de Buenos Aires, Argentina. Revista del Museo Argentina Ciencias Naturales "Bernandino Rivadavia" e Instituto Nacional de Investigacion de la Ciencias Naturales. *Zoologica* 10:27–63.

Gatz, A. J., Jr. 1981. Non-random mating by size in American toads, *Bufo americanus*. *Anim. Behav.* 29:1004–12.

Gerhardt, H. C. 1974. The significance of some spectral features in mating call recognition in the green treefrog (*Hyla cinerea*). *J. Exp. Biol.* 74:59–73.

———. 1975. Sound pressure levels and radiation patterns of the vocalizations of some North American frogs and toads. *J. Comp. Physiol.* 102:1–12.

———. 1978a. Mating call recognition in the green treefrog (*Hyla cinerea*): The significance of some fine-temporal properties. *J. Exp. Biol.* 74:59–73.

———. 1978b. Temperature coupling in the vocal communication system of the gray treefrog, *Hyla versicolor*. *Science* 199:992–94.

———. 1981a. Mating call recognition in the green treefrog (*Hyla cinerea*):

Importance of two frequency bands as a function of sound pressure level. *J. Comp. Physiol.* 144:9–15.

———. 1981b. Mating call recognition in the barking treefrog (*Hyla gratiosa*): Responses to synthetic calls and comparisons with the green treefrog (*Hyla cinerea*). *J. Comp. Physiol.* 144:17–25.

———. 1982. Sound pattern recognition in some North American treefrogs (Hylidae): Implications for mate choice. *Amer. Zool.* 22:581–95.

———. 1983. Communication and the environment. In *Animal behaviour, 2: Communication,* ed. T. R. Halliday and P. J. B. Slater. New York: W. H. Freeman and Co.

Gerhardt, H. C., and Rheinlaender, J. 1980. Accuracy of sound localization in a miniature dendrobatid frog. *Naturwissenschaften* 67:362–63.

Gish, S. L., and Morton, E. S. 1981. Structural adaptations to local habitat acoustics in Carolina wren songs. *Z. Tierpsychol.* 56:74–84.

Gould, S. J. 1974. The origin and function of "bizarre" structures: Antler size and skull size in the "Irish elk," *Megalocerus giganteus. Evolution* 28:191–220.

Greenwalt, C. H. 1968. *Bird song: Acoustics and physiology.* Washington, D.C.: Smithsonian Institution Press.

Greer, B. J., and Wells, K. D. 1980. Territorial and reproductive behavior in a centrolenid frog *Centrolenella fleischmanni. Herpetologica* 36:318–26.

Griffin, D. R. 1958. *Listening in the dark.* New Haven: Yale Univ. Press.

Haas, R. 1976. Sexual selection in *Nothobranchius guentheri* (Pisces: Cyprinodontidae). *Evolution* 30:614–22.

Halliday, T. R. 1978. Sexual selection and mate choice. In *Behavioural ecology: An evolutionary approach,* ed. J. R. Krebs and N. B. Davies. Sunderland, Mass.: Sinauer Assoc.

———. 1983a. Do frogs and toads choose their mates? *Nature* 306:226–27.

———. 1983b. The study of mate choice. In *Mate choice,* ed. P. Bateson. Cambridge: Cambridge Univ. Press.

Hamilton, W. D. 1971. Geometry for the selfish herd. *J. Theor. Biol.* 31:295–311.

Hamilton, W. D., and Zuk, M. Z. 1982. Heritable true fitness and bright birds: A role for parasites? *Science* 218:384–87.

Hausfater, G. 1975. Dominance and reproduction in the baboon (*Papio cynocephalus*). *Cont. to Primatology* 3:1–150.

Heffner, R., and Heffner, H. 1980. Hearing in the elephant (*Elaphas maximus*). *Science* 208:518–20.

Heyer, W. R. 1969. The adaptive ecology of the species groups of the genus *Leptodactylus. Evolution* 23:421–28.

———. 1975. A preliminary analysis of the intergeneric relationships of the frog family Leptodactylidae. *Smithsonian Contrib. to Zool.,* no. 199.

———. 1980. Personal communication with author.

Heyer, W. R., and Rand, A. S. 1977. Foam nest construction in the leptodactylid frogs, *Leptodactylus pentadactylus* and *Physalaemus pustulosus* (Amphibia, Anura, Leptodactylidae). *J. Herpetol.* 11:225–28.

Hillman, S. S., and Withers, P. C. 1979. An analysis of respiratory surface area as a limit to activity metabolism in anurans. *Can. J. Zool.* 57:2100–05.

Hinde, R. A. 1970. *Animal behaviour: A synthesis of ethology and comparative psychology.* New York: McGraw-Hill.

Hogan-Warburg, A. J. 1966. Social behavior of the ruff *Philomachus pugnax* (L.) *Ardea* 54:109–229.

Howard, R. D. 1978a. The evolution of mating strategies in bullfrogs, *Rana catesbeiana. Evolution* 32:850–71.

―――. 1978b. The influence of male defended oviposition sites on early embryo mortality in bullfrogs. *Ecology* 59:789–98.

―――. 1979. Estimating reproductive success in natural populations. *Amer. Natur.* 114:221–31.

―――. 1980. Mating behaviour and mating success in woodfrogs, *Rana sylvatica. Anim. Behav.* 28:705–16.

―――. 1981. Sexual dimorphism in bullfrogs. *Ecology* 62:303–10.

―――. 1983. Sexual selection and variation in reproductive success. *Amer. Natur.* 122:301–25.

Hoy, R. R., and Paul, R. C. 1973. Genetic control of song specificity in crickets. *Science* 180:82–83.

Huxley, J. S. 1938a. Darwin's theory of sexual selection and the data subsumed by it, in the light of recent research. *Amer. Natur.* 72:416–33.

―――. 1938b. The present standing of the theory of sexual selection. In *Evolution: Essays on aspects of evolutionary biology presented to Professor E. S. Goodrich on his seventieth birthday.* Oxford: Clarendon Press.

Jaeger, R. G. 1976. A possible prey-call window in anuran auditory processing. *Copeia* 1976:833–34.

Jameson, D. L. 1955. Evolutionary trends in courtship and mating behavior of salientia. *Syst. Zool.* 4:105–19.

Janetos, A. C. 1980. Strategies of female choice: A theoretical analysis. *Behav. Ecol. Sociobiol.* 7:107–12.

Jenni, D. A. 1974. Evolution of polyandry in birds. *Amer. Zool.* 14:129–44.

Karr, J. R. 1982. Avian extinction on Barro Colorado Island, Panama: A reassessment. *Amer. Natur.* 119:220–39.

Kirkpatrick, M. 1982. Sexual selection and the evolution of female choice. *Evolution* 36:1–12.

Kluge, A. G. 1981. The life history, social organization, and parental behavior of *Hyla rosenbergi* Boulenger, a nest-building gladiator frog. *Misc. Publ. Mus. Zool. Univ. Michigan*, no. 160.

Knudsen, E. I. 1980. Sound localization in birds. In *Comparative studies of hearing in vertebrates*, ed. A. N. Popper and R. R. Fay. New York: Springer-Verlag.

Knudsen, E. I., and Konishi, M. 1979. Mechanism of sound localization by the barn own *(Tyto alba). J. Comp. Physiol.* 133:13–21.

Kruijt, J. P., and Hogan, J. A. 1967. Social behaviour on the lek in black grouse, *Lyrurus tetrix tetrix* (L.) *Ardea* 55:204–40.

Kruse, K. C., and Mounce, M. 1982. The effects of multiple matings of fertilization capability in male American toads *(Bufo americanus). J. Herpetol.* 16:410–11.

Lande, R. 1976. The maintenance of genetic variability by mutation in a polygenic character with linked loci. *Genet. Res. Camb.* 26:221–35.

———. 1980. Sexual dimorphism, sexual selection, and adaptation in polygenic characters. *Evolution* 34:292–305.

———. 1981. Models of speciation by sexual selection on polygenic characters. *Proc. Nat. Acad. Sci.* 78:3721–25.

———. 1982. Rapid origin of sexual isolation and character divergence in a cline. *Evolution* 36:213–23.

Lawick-Goodall, J. 1971. *In the shadow of man.* Boston: Houghton Mifflin.

Lenington, S. 1980. Female choice and polygyny in redwinged blackbirds. *Anim. Behav.* 28:347–61.

Lewis, B., and Coles, R. 1980. Sound localization in birds. *Trends in Neurosci.* 3:102–5.

Lewontin, R. C. 1970. *The genetic basis of evolutionary change.* New York: Columbia Univ. Press.

———. 1977. Fitness, survival, and optimality. In *Analysis of ecological systems,* ed. D. J. Horn, G. R. Stoirs, and R. D. Mitchell. Columbus: Ohio State Univ. Press.

Licht, L. E. 1976. Sexual selection in the toad, *Bufo americanus. Can. J. Zool.* 54:1277–84.

Littlejohn, M. J. 1958. Mating behavior in the treefrog *Hyla versicolor. Copeia* 1958:222–23.

———. 1959. Call differences in a complex of seven species of *Crinia* (Anura, Leptodactylidae). *Evolution* 13:452–68.

———. 1960. Call discrimination and potential reproductive isolation in *Pseudacris triseriata* females from Oklahoma. *Copeia* 1960:370–71.

———. 1965. Premating isolating mechanisms in the *Hyla ewingi* complex (Anura: Hylidae). *Evolution* 19:234–43.

———. 1969. The systematic significance of isolating mechanisms. In *Systematic biology,* publ. 1692. Washington, D.C.: National Academy of Sciences.

Littlejohn, M. J., and Loftus-Hills, J. J. 1968. An experimental evaluation of premating isolation in the *Hyla ewingi* complex. *Evolution* 22:659–63.

Loftus-Hills, J. J., and M. J. Littlejohn. 1971. Pulse repetition rate as the basis for mating call discrimination by two sympatric species of *Hyla. Copeia* 1971:154–56.

Loftus-Hills, J. J., and Michaud, T. 1952. Mating call discrimination by females of Strecker's chorus frog (*Pseudacris streckeri*). *Texas J. Science* 11:86–92.

Lorenz, K. L. 1981. *The foundations of ethology.* New York: Springer-Verlag.

Lutz, B. 1960. Fighting and an incipient notion of territoriality in male tree frogs. *Copeia* 1960:61–63.

Lynch, J. D. 1968. Genera of leptodactylid frogs in Mexico. *Univ. Kansas Mus. Nat. Hist. Pub.* 17:503–15.

———. 1970. Systematic status of the American leptodactylid genera *Engystomops, Euphemix,* and *Physalaemus. Copeia* 1970:488–96.

———. 1971. Evolutionary relationships, osteology, and zoogeography of the American leptodactylid frogs. *Univ. Kansas Mus. Nat. Hist. Pub.*, no. 53.

McClanahan, L. L.; Stinner, J. N.; and Shoemaker, V. H. 1971. Skin lipids, water loss, and energy metabolism in a South American tree frog (*Phyllomedusa sauvegi*). *Physiol. Zool.* 51:179–87.

MacDonald, J., and Crossley, S. 1982. Behavioural analysis of lines selected for wing vibration in *Drosophila melanogaster*. *Anim. Behav.* 30:802–10.

MacNally, R. C. 1981. On the reproductive energetics of chorusing males: Energy depletion, restoration, and growth for two sympatric species of *Ranidella* (Anura). *Oecologia* 51:181–88.

MacNally, R. C., and Young, D. 1981. Song energetics of the bladder cicada, *Cystosoma saundersii*. *J. Exp. Biol.* 90:185–96.

Maier, V. 1982. Acoustic communication in the guinea fowl (*Numida meleagris*): Structure and the use of vocalizations, and the principle of message coding. *Z. Tierpsychol.* 59:29–83.

Marler, P. 1955. Characteristics of some animal calls. *Nature* 176:6–8.

Marler, P., and Hamilton, W. J., III. 1966. Mechanisms of animal behavior. New York: John Wiley and Sons.

Marten, K., and Marler, P. 1977. Sound transmission and its significance for animal vocalization, 1: Temperate habitats. *Behav. Ecol. Sociobiol.* 2:271–90.

Marten, K.; Quine, D.; and Marler, P. 1977. Sound transmission and its significance for animal vocalization, 2: Tropical habitats. *Behav. Ecol. Sociobiol.* 2:291–302.

Martin, A. A. 1970. Parallel evolution in the adaptive ecology of leptodactylid frogs of South America and Australia. *Evolution* 24:643–44.

Martin, W. F. 1972. Evolution of vocalizations in the genus *Bufo*. In *Evolution in the genus Bufo*, ed. W. F. Blair. Austin: Univ. Texas Press.

Martin, W. F., and Gans, C. 1972. Muscular control of the vocal tract during release signalling in the toad *Bufo valliceps*. *J. Morphol.* 137:1–27.

Martof, B. S. 1953. Territoriality in the green frog, *Rana clamitans*. *Ecology* 34:165–75.

Maynard Smith, J. 1956. Fertility, mating behavior, and sexual selection in *Drosophila subobscura*. *J. Genetics* 54:261–79.

———. 1976. Sexual selection and the handicap principle. *J. Theor. Biol.* 57:239–42.

———. 1978. *The evolution of sex*. Cambridge: Cambridge Univ. Press.

Mayr, E. 1963. *Animal species and evolution*. Cambridge: Harvard Univ. Press.

———. 1972. Sexual selection and natural selection. In *Sexual selection and the descent of man*, ed. B. Campbell. Chicago: Aldine.

———. 1982. *The growth of biological thought*. Cambridge: Harvard Univ. Press.

———. 1983. How to carry out the adaptationist program? *Amer. Natur.* 121:324–34.

Mecham, J. S. 1961. Isolating mechanisms in anuran amphibians. In *Vertebrate speciation*, ed. W. F. Blair. Austin: Univ. Texas Press.

Michaud, T. C. 1962. Call discrimination by females of the chorus frogs, *Pseudacris clarki* and *Pseudacris nigrata*. *Copeia* 1962:213–15.

Michelson, A. 1978. Sound reception in different environments. In *Sensory ecology, review, and perspectives*, ed. M. Ali. New York: Plenum Pub.
Milstead, W. W. 1960. Frogs of the genus *Physalaemus* in Brazil with a description of a new species. *Copeia* 1960:83–89.
Moffat, A. J. M., and Capranica, R. R. 1974. Sensory processing in the peripheral auditory system of treefrogs (*Hyla*) *J. Acoust. Soc. Amer.* 55:480.
Morton, E. S. 1975. Ecological sources of selection on avian sounds. *Amer. Natur.* 109:17–34.
———. 1977. On the occurrence of motivational structural rules in some bird and mammal sounds. *Amer. Natur.* 111:855–69.
Moynihan, M. 1970. Control, suppression, decay, disappearance, and replacement of displays. *J. Theor. Biol.* 29:85–112.
Mudry, K. M., and Capranica, R. R. 1980. Evoked auditory activity within the telencephalon of the bullfrog (*Rana catesbeiana*). *Brain Research* 182:303–11.
Mudry, K. M.; Constantine-Paton, M.; and Capranica, R. R. 1977. Auditory sensitivity of the diencephalon of the leopard frog, *Rana p. pipiens*. *J. Comp. Physiol.* 114:1–13.
Nagy, K. A. 1983. Ecological energetics of a lizard. In *Lizard ecology: Studies of a model organism*, ed. R. B. Huey, E. R. Pianka, and T. W. Schoener. Cambridge: Harvard Univ. Press.
Narins, P. M. 1983. Responses to the torus semicircularis of the coqui treefrog to FM sinusoids. In *Advances in vertebrate neuroethology*, ed. J. P. Ewert, R. R. Capranica, and D. J. Ingle, New York: Plenum Pub.
Narins, P. M., and Capranica, R. R. 1976. Sexual differences in the auditory system of the treefrog, *Eleutherodactylus coqui*. *Science* 192:378–80.
———. 1978. Communicative significance of the two-note call of the treefrog, *Eleutherodactylus coqui*. *J. Comp. Physiol.* 127:1–9.
———. 1980. Neural adaptations for processing two-note calls of the Puerto Rican treefrog, *Eleutherodactylus coqui*. *Brain Behav. Evol.* 17:48–66.
Nice, M. M. 1941. The role of territory in the life of birds. *Amer. Midl. Natur.* 26:441–87.
Noble, G. K. [1931] 1954. *The biology of the amphibia*. New York: Dover Press.
Noble, G. K., and Aronson, L. R. 1942. The sexual behavior of Anura, 1: The normal mating pattern of *Rana pipiens*. *Bull. Amer. Mus. Nat. Hist.* 80:127–42.
Noble, G. K., and Bradley, H. T. 1933. The mating behavior of lizards: Its bearing on the theory of sexual selection. *N. Y. Acad. Sci.* 35:25–100.
O'Donald, P. 1962. The theory of sexual selection. *Heredity* 17:541–52.
———. 1967. A general model of sexual and natural selection. *Heredity* 22:499–518.
———. 1972. Sexual selection by variations in fitness at breeding time. *Nature* 237:349–51.
———. 1973. Models of sexual and natural selection in polygynous species. *Heredity* 31:145–56.
———. 1978. Theoretical aspects of sexual selection: A generalized model of mating behavior. *Theor. Pop. Biol.* 13:226–43.

———. 1980a. *Genetic models of sexual selection.* Cambridge: Cambridge Univ. Press.
———. 1980b. Sexual selection by female choice in a monogamous bird: Darwin's theory corroborated. *Heredity* 45:210–17.
———. 1983. Sexual selection by female choice. In *Mate choice.*, ed. P. Bateson. Cambridge: Cambridge Univ. Press.
Oldham, R. S., and Gerhardt, H. C. 1975. Behavioral isolating mechanisms of the treefrogs *Hyla cinerea* and *Hyla gratiosa. Copeia* 1975:223–31.
Orians, G. H. 1969. On the evolution of mating systems in birds and mammals. *Amer. Natur.* 103:589–603.
O'Rourke, F. J. 1970. *The fauna of Ireland.* Cork: Mercier Press.
Oster, G. F., and Wilson, E. O. 1978. *Caste and ecology in social insects.* Monogr. in Pop. Biol., no. 12. Princeton: Princeton Univ. Press.
Otte, D. 1979. Historical development of sexual selection. In *Sexual selection and reproductive competition in insects*, ed. M. S. Blum and N. H. Blum. New York: Academic Press.
Parker, G. A. 1982. Phenotype limited evolutionary stable strategies. In *Current problems in sociobiology,* ed. Kings College Sociobiology Group. Cambridge: Cambridge Univ. Press.
———. 1983. Mate choice, sexual selection, mate quality, and mating success. In *Mate choice,* ed. P. Bateson. Cambridge: Cambridge Univ. Press.
Partridge, L. 1980. Mate choice increases a component of offspring fitness in fruit flies. *Nature* 283:290–91.
———. 1983. Non-random mating and offspring fitness. In *Mate choice,* ed. P. Bateson. Cambridge: Cambridge Univ. Press.
Passmore, N. I. 1981. Sound level of mating calls of some African frogs. *Herpetologica* 37:166–71.
Paterson, H. E. H. 1982. Perspectives on speciation by reinforcement. *So. African J. Sci.* 78:53–57.
Payne, R. B., and Payne, K. 1977. Social organization and mating success in local song populations of village indigobirds, *Vidua chalybeata. Z. Tierpsychol.* 45:113–73.
Perrill, S. A.; Gerhardt, H. C.; and Daniels, R. 1978. Sexual parasitism in the green treefrog *(Hyla cinerea). Science* 200:1179–80.
———. 1982. Mating strategy shifts in male green treefrogs (*Hyla cinerea*): an experimental approach. *Anim. Behav.* 30:43–48.
Pianka, E. R., and Parker, W. S. 1975. Age-specific reproductive tactics. *Amer. Natur.* 109:453–64.
Pleszczynska, W. K. 1978. Microgeographic prediction of polygyny in the lark bunting. *Science* 201:935–37.
Pough, F. H. 1980. The advantages of ectothermy for tetrapods. *Amer. Natur.* 115:92–112.
Ramer, J. D.; Jennson, T. A.; and Hurst, C. J. 1983. Size-related variation in the advertisement call of *Rana clamitans* (Anura: Ranidae), and its effect on conspecific males. *Copeia* 1983:141–55.
Rand, A. S. 1976. Letter to author. 31 March.
Rand, A. S., and Rand, W. M. 1982. Variation in rainfall on Barro Colorado

Island. In *The ecology of a tropical rainforest: Seasonal rhythms and long term changes*, ed. E. G. Leigh, Jr., A. S. Rand, and D. M. Windsor. Washington, D.C.: Smithsonian Institution Press.

Rand, A. S., and Ryan, M. J. 1981. The adaptive significance of a complex vocal repertoire in a Neotropical frog. *Z. Tierpsychol.* 57:209–14.

Rand, W. M., and Rand, A. S. 1976. Agonistic behavior in nesting iguanas: A stochastic analysis of dispute settlement dominated by the minimization of energy cost. *Z. Tierpsychol.* 40:279–99.

Richards, D. G. 1981. Estimation of distance of singing conspecifics by the Carolina wren, *Auk* 98:127–33.

Rivero, J. A. 1978. *Los anfibios y reptiles de Puerto Rico.* San Juan: Edit. Univ. Puerto Rico.

Rivero, J. A., and Esteves, A. E. 1969. Observations on the agonistic and breeding behavior of *Leptodactylus pentadactylus* and other amphibian species of Venezuela. *Breviora* 321:1–14.

Roeder, K. D. 1962. The behaviour of free flying moths in the presence of artificial ultrasonic pulses. *Anim. Behav.* 10:300–304.

Roederer, J. G. 1975. *Introduction to the physics and psychophysics of music.* New York: Springer-Verlag.

Rossing, T. D. 1982. *The science of sound.* Reading, Mass.: Addison-Wesley Pub.

Ruse, M. 1979. *The Darwinian revolution: Science red in tooth and claw.* Chicago: Univ. Chicago Press.

Ryan, M. J. 1980a. Female mate choice in a Neotropical frog. *Science* 209:523–25.

―――. 1980b. The reproductive behavior of the bullfrog, *Rana catesbeiana. Copeia* 1980:108–14.

―――. 1983a. Frequency modulated calls and species recognition in a Neotropical frog, *Physalaemus pustulosus. J. Comp. Physiol.* 150:217–21.

―――. 1983b. Sexual selection and communication in a Neotropical frog, *Physalaemus pustulosus. Evolution* 37:261–72.

―――. 1984. Energetic efficiency of calling in the frog *Physalaemus pustulosus. J. Exp. Biol.*

Ryan, M. J.; Bartholomew, G. A.; and Rand, A. S. 1983. Energetics of reproduction in a Neotropical frog, *Physalaemus pustulosus. Ecology,* 64:1456–62.

Ryan, M. J., and Brenowitz, E. H. 1984. The role of body size, phylogeny, and ambient noise in the evolution of bird song. *Amer. Natur.*

Ryan, M. J., and Tuttle, M. D. 1983. The ability of the frog-eating bat to distinguish among potentially poisonous prey items using acoustic cues. *Anim. Behav.* 37:827–33.

Ryan, M. J.; Tuttle, M. D.; and Barclay, R. M. R. 1983. Behavioral responses of the frog-eating bat to sonic frequencies. *J. Comp. Physiol.* 150:413–18.

Ryan, M. J.; Tuttle, M. D.; and Rand, A. S. 1982. Bat predation and sexual advertisement in a Neotropical frog. *Amer. Natur.* 119:136–39.

Ryan, M. J.; Tuttle, M. D.; and Taft, L. K. 1981. The costs and benefits of frog chorusing behavior. *Behav. Ecol. Sociobiol.* 8:273–78.

Salthe, S. N., and Duellman, W. E. 1973. Quantitative constraints associated with reproductive mode in anurans. In *The evolutionary biology of anurans*, ed. J. L. Vial. Columbia: Univ. Missouri Press.

Salthe, S. N., and Mecham, J. S. 1974. Reproductive and courtship patterns. In *Physiology of the amphibia 2*, ed. B. Lofts. New York: Academic Press.

Savage, J. M. 1966. The origins and history of the Central American herpetofauna. *Copeia* 1966:719–65.

———. 1973. The geographic distribution of frogs: Patterns and predictions. In *The evolutionary biology of anurans*, ed. J. L. Vial. Columbia: Univ. Missouri Press.

———. 1982. The enigma of the Central American herpetofauna: Dispersals or vicariance? *Ann. Missouri Bot. Garden* 69:464–547.

Schoener, T. W., and Schoener, A. 1982. The ecological correlates of survival in some Bahamian anolis lizards. *Oikos* 39:1–16.

Searcy, W. A. 1979. Female choice of mates: A general model for birds and its application to red-winged blackbirds (*Aeglaius phoeniceus*). *Amer. Natur.* 114:77–100.

———. 1982. The evolutionary effects of mate selection. *Ann. Rev. Ecol. Syst.* 13:57–85.

Selander, R. K. 1965. On mating systems and sexual selection. *Amer. Natur.* 99:129–41.

———. 1972. Sexual selection and dimorphism in birds. In *Sexual selection and the descent of man*, ed. B. Campbell. Chicago: Aldine.

Sexton, O. J., and Ortleb, E. P. 1966. Cues used by the leptodactylid frog, *Physalaemus pustulosus*, in the selection of an oviposition site. *Copeia* 1966:226–30.

Seymour, R. S. 1973. Physiological correlates of forced activity and burrowing in the spadefoot toad, *Scaphiopus hammondi*. *Copeia* 1973:103–5.

Shalter, M. D. 1978. Localization of passerine seets and mobbing calls by goshawks and pygmy owls. *Z. Tierpsychol.* 46:260–67.

Shields, W. M. 1982. *Inbreeding, philopatry, and the evolution of sex*. Albany: State Univ. New York Press.

Shine, R. 1979. Sexual selection and sexual dimorphism in the Amphibia. *Copeia* 1979:297–306.

Siegel, S. 1956. *Nonparametric statistics for the behavioral sciences*. New York: McGraw-Hill.

Smith, G. C. 1976. Ecological energetics of three species of ectothermic vertebrates. *Ecology* 57:252–64.

Smith, W. J. 1977. *The behavior of communicating: An ethological approach*. Cambridge: Harvard Univ. Press.

Smith-Gill, S. J., and Berven, K. A. 1980. In vitro fertilization and assessment of male reproductive potential using mammalian gonadotropin-releasing hormone to induce spermiation in *Rana sylvatica*. *Copeia* 1980:723–28.

Sokal, R. R., and Rohlf, F. J. 1969. *Biometry: The principles and practice of statistics in biological sciences*. San Francisco: W. H. Freeman and Co.

Stearns, S. C. 1976. Life history tactics: A review of the ideas. *Quart. Rev. Biol.* 51:3–47.

———. 1980. A new view of life-history evolution. *Oikos* 35:266–81.

Steele, R. G. D., and Torrie, J. H. 1960. *Principles and procedures of statistics with special reference to the biological sciences.* New York: McGraw-Hill.

Straughn, I. R., and Heyer, W. R. 1976. A functional analysis of the mating calls of the Neotropical frog genera of the *Leptodactylus* complex (Amphibia, Leptodactylidae). *Papeis Avulsos de Zoologica (Sao Paulo)* 29:221–45.

Suga, N. 1978. Specializations of the auditory system for reception and processing of species-specific sounds. *Fed. Proc.* 37:2342–54.

Suga, N., and Schlegel, P. 1973. Coding and processing in the auditory systems of FM signal-producing bats. *J. Acoust. Soc. Amer.* 54:174–90.

Sullivan, B. K. 1982a. Sexual selection in woodhouse's toad (*Bufo woodhousei*), 1: Chorus organization. *Anim. Behav.* 30:680–86.

———. 1982b. Significance of size, temperature, and call attributes to sexual selection in *Bufo woodhousei australis. J. Herpetol.* 16:103–6.

———. 1983. Sexual selection in the Great Plains toad (*Bufo cognatus*). *Behaviour* 84:258–64.

Taigen. T. L.; Emerson, S. B.; and Pough, F. H. 1982. Ecological correlates of anuran exercise physiology, *Oecologia* 52:49–56.

Tannenbaum, B. R. 1974. Reproductive strategies in a tropical bat. Ph.D. diss., Cornell Univ.

Taylor, D. H., and Guttman, S. I., eds. 1977. *The reproductive biology of amphibians.* New York: Plenum Pub.

Thornhill, R. 1980. Competitive, charming males and chosy females: Was Darwin correct? *Fla. Entomol. Soc.* 63:5–30.

Tinkle, D. W., and Hadley, N. F. 1975. Lizard reproduction: Caloric estimates and comments on its evolution. *Ecology* 56:427–34.

Tinkle, D. W.; Wilbur, H. W.; and Tilley, S. G. 1969. Evolutionary strategies in lizard reproduction. *Evolution* 24:55–74.

Toft, C. A. 1981. Feeding ecology of Panamanian litter anurans. *J. Herpetol.* 15:139–44.

Townsend, D. S.; Stewart, M. M.; Pough, F. H.; and Brussard, P. F. 1981. Internal fertilization in an oviparous frog. *Science* 212:469–71.

Trivers, R. L. 1972. Parental investment and sexual selection. In *Sexual selection and the descent of man*, ed. B. Campbell. Chicago: Aldine.

———. 1976. Sexual selection and resource-accrual abilities in *Anolis garmani. Evolution* 30:253–69.

Tuttle, M. D., and Ryan, M. J. 1981. Bat predation and the evolution of frog vocalizations in the Neotropics. *Science* 214:677–78.

Tuttle, M. D.; Taft, L. K.; and Ryan, M. J. 1982. Evasive behaviour of a frog in response to bat predation. *Anim. Behav.* 30:393–97.

Vial, J. L., ed. 1973. *Evolutionary biology of the anurans.* Columbia: Univ. Missouri Press.

Villa, J. 1972. *Anfibios de Nicaragua.* Managua: Instituto Geografica y Nacional Banco Central de Nicaragua.

Wade, M. J. 1979. Sexual selection and variance in reproductive success. *Amer. Natur.* 114:742–46.

Wade, M. J., and Arnold, S. J. 1980. The intensity of sexual selection in relation to male sexual behavior, female choice, and sperm precedence. *Anim. Behav.* 28:446–61.

Waldman, B. 1980. Can a female toad choose her mate? Typescript.

Walkowiak, W. 1980. The coding of auditory signals in the torus semicircularis of fire-bellied toads and the grass frog: Responses to simple stimuli and to conspecific calls. *J. Comp. Physiol.* 138:131–48.

Wallace, A. R. 1905. *Darwinism*. 3d ed. London: Macmillan.

Waser, P. M., and Waser, M. S. 1977. Experimental studies of primate vocalizations: Specializations for long-distance propagation. *Z. Tierpsychol.* 43:239–63.

Weathers, W. W., and Snyder, G. K. 1977. Relation of oxygen consumption to temperature and time of day in tropical anuran amphibians. *Aust. J. Zool.* 25:19–24.

Weismann, A. 1904. *The evolutionary theory*. London: E. Arnold Co.

Wells, K. D. 1977a. The social behaviour of anuran amphibians. *Anim. Behav.* 25:666–93.

──────. 1977b. Territoriality and male mating success in the green frog (*Rana clamitans*). *Ecology* 58:750–62.

Wells, K. D., and Schwartz, J. J. 1982. The effect of vegetation on the propagation of calls in the Neotropical frog *Centrolenella fleischmanni*. *Herpetologica* 38:449–55.

West Eberhard, M. J. 1979. Sexual selection, social competition, and evolution. *Proc. Amer. Phil. Soc.* 123:222–34.

──────. 1983. Sexual selection, social competition, and speciation. *Quart. Rev. Biol.* 58:155–83.

Whitney, C. L., and Krebs, J. R. 1975. Mate selection in Pacific tree frogs. *Nature* 255:325–26.

Wiewandt, T. 1969. Vocalizations, aggressive behavior, and territoriality in the bullfrog, *Rana catesbeiana*. *Copeia* 1969:276–85.

Wilbur, H. M.; Rubenstein, D. I.; and Fairchild, L. 1978. Sexual selection in toads: The role of female choice and male body size. *Evolution* 32:264–70.

Wilczynski, W.; Resler, C.; and Capranica, R. R. 1981. A study of the mechanism underlying the directional sensitivity of the anuran ear *Neurosci. Abstr.* 7:147.

Wiley, E. O. 1981. *Phylogenetics: The theory and practice of phylogenetic systematics*. New York: John Wiley and Sons.

Wiley, R. H. 1974. Evolution of social organization and life-history patterns among grouse. *Quart. Rev. Biol.* 49:201–27.

Wiley, R. H., and Richards, D. G. 1978. Physical contraints on acoustical communication in the atmosphere: Implications for the evolution of animal vocalizations. *Behav. Ecol. Sociobiol.* 3:69–94.

──────. 1982. Adaptations for acoustic communication in birds: Sound transmission and signal detection. In *Acoustic communication in birds*,

1: Production, propagation, and design features, ed. D. E. Kroodsma and E. H. Miller. New York: Academic Press.

Williams, G. C. 1966. *Adaptation and natural selection: A critique of some current evolutionary thought*. Princeton: Princeton Univ. Press.

―――. 1975. *Sex and evolution*. Princeton: Princeton Univ. Press.

Wilson, E. O. 1975. *Sociobiology: The new synthesis*. Cambridge: Harvard Univ. Press.

Wittenberger, J. F. 1978. The evolution of mating systems in grouse. *Condor* 80:126–37.

Woolbright, L. L. 1983. Sexual selection and size dimorphism in anurans. *Amer. Natur.* 121:110–19.

Wrangham, R. W. 1980. Female choice of least costly males: A possible factor in the evolution of leks. *Z. Tierpsychol.* 54:357–67.

Zahavi, A. 1975. Mate selection: A selection for a handicap. *J. Theor. Biol.* 53:205–14.

―――. 1977. The cost of honesty (further remarks on the handicap principle). *J. Theor. Biol.* 67:603–5.

Zelick, R. D., and Narins, P. M. 1982. Analysis of acoustically evoked call suppression behaviour in a Neotropical treefrog. *Anim. Behav.* 30:728–33.

Zweifel, R. G. 1968. Effects of temperature, body size, and hybridization on mating calls of toads, *Bufo a. americanus* and *Bufo woodhousii fowleri*. *Copeia* 1968:269–85.

Index

Acoustically foraging predators. *See* Predation, based on acoustic cues

Acoustic processing. *See* Auditory system of anurans

Acoustic recording and analysis, methods of, 193–94

Acoustic stimuli, preparation of, 194

Acris crepitans, 21

Additive genetic variation, 129. *See also* Heritability

Adenomera, 30

Adenomera hylaedactyl, 98

Adenomera marmorata, 98

Advertisement call: complexity series of, 36–37, 66–71; definition of, 20; description of, 36, 66–71; differences among populations, 21–22; differences based on size, 22, 87–92, 96–104; effect of sexual selection on the evolution of, 96–104; energetic efficiency in the production of, 148–50; energy expended during the production of, 75–76, 144–48; frequency modulation of, 67–70, 108, 111–20; function of, 66–122; function of whine and chuck components of, 107–11; harmonic structure of 67–72, 116–20, 124–26; influence of the environment on the transmission of, 99, 114–20; influence of male interactions on, 23, 66, 73–75, 95–96; influence of structure on localizability of, 77–78, 83–87; influence of structure on predation, 77–84, 164–65; intensity of, 39, 60, 75, 92, 149, 164–65; morphological correlate of the evolution of, 104–7; numbers of produced each night, 39–41, 59–60, 144–46; production of, 71–73; relative attenuation of components of, 114–20; repetition rate of, 39, 60, 145–47; role of in anuran courtship reviewed, 15–16; role of in species recognition, 20, 23, 31, 66, 107–11, 113–14, 186–87; sensory processing of, *see* Auditory system, of anurans; social mediation of complexity of, 73–75; as a species-isolating mechanism, *see* Role in species recognition; spectral characteristics of, 66–71, 87–104, 105–8, 111–20, 124–26; temporal characteristics of, 66–71; total power of, 148–50; transmission properties of, 22, 114–20, 149–50

Aerobic capacity, comparison of among anurans, 152

Aerobic dependence, 154–55; index of, 154

Aerobic metabolic scope, 151

Aerobic metabolism: *See* Energy expended; Oxygen consumption

Agalychnis callidryas, 25, 27, 47

Alarm calls, structure of, 77–78

Alexander, R. D., 59, 172, 176

Alternative male reproductive behaviors, 182–84

Altig, R., 7, 67

Amphibian papilla, 20, 112, 124. *See also* Auditory system of anurans

Amplexus, 13, 44; displacements of male during, 17, 44

Anaerobic metabolism. *See* Lactate; Energy expended

Anderrson, M., 9, 128, 130, 142

Andrews, R. M., 144

Index

Anolis, 144
Anuran breeding patterns, review of, 13–15
Apomorphic, 102
Arak, A., 19, 23
Argus pheasant, 3
Arnold, S. J., 8, 9, 14, 61, 94, 123, 143
Aronson, L. R., 42
Assortative mating: in anurans, 132; lack of in *P. pustulosus*, 137–42
Auditory system of anurans, 16, 20–21, 92, 111, 124, 138
Auditory thalamic region, 21
Auditory tuning curve, 20
Axillary amplexus, 29
Axtell, R. W., 20

Baboons, 15
Baker, R. R., 4
Barclay, R. M. R., 83
Barrio, A., 30–31, 73, 99, 107–9
Barro Colorado Island, patterns of rainfall on, 32–33
Bartholomew, G. A., 76, 144–49, 153, 158, 160, 179, 199
Basilar papilla, 20, 112, 124–25. *See also* Auditory system of anurans
Bateman, A. J., 8
Bateson, P., 2
Bennett, A. F., 151, 153–54
Beranek, L. L., 105, 114
Berven, K. A., 17, 132, 136, 182
Binaural comparisons as a mechanism for sound localization, 77, 84
Binaural cues used for sound localization, 77
Bioacoustic techniques. *See* Acoustic recording and analysis, methods of
Bird song, production of, 71
Blair, W. F., 13, 16, 20–21, 100
Bogert, C. M., 16, 19
Borgia, G. 138
Bowman, R. I., 114, 117
Brackenbury, J. H., 71, 149
Bradbury, J. W., 10, 59, 172, 177
Bradley, H. T., 5–6, 12
Brattstrom, B. H., 19, 31
Bray, D. E., 15, 49
Breeding season of *P. pustulosus*, 32–34
Breeding sites of *P. pustulosus*, 33
Brenowitz, E. A., 114–15, 150
Brown, C., 77–78, 81
Brown, J. L., 10, 143
Brown, L., 93

Bucher, T. L., 76, 144–47, 153, 199
Buechner, H. K., 10
Bufo americanus, 16–17, 60, 62, 132, 136, 184
Bufo bufo, 16, 23, 44, 62, 132, 136
Bufo canorus, 62
Bufo cognatus, 17, 184
Bufo exsul, 62
Bufo fowleri, 17
Bufo marinus, 164, 175
Bufo quercicus, 16
Bufo speciosus, 20
Bufo typhonius, 41–42, 62
Bullfrogs, 18, 42, 49, 124, 126, 131, 138, 187
Burger, 97

Cade, W. H., 19, 97, 129, 164
Call frequency as a function of male size. *See* Advertisement call, differences based on size
Calling, temporal patterns in, in *P. pustulosus*, 35–44
Campbell, B., 1, 10
Cannatella, D. C., 109
Capranica, R. R., 16, 20–22, 84, 93, 105, 111–13, 124, 138
Carey, C., 145
Cat-eyed snakes, 182
Cei, J., 99
Central nervous system, auditory processing in, 20–21, 112
Centrolenella, 63
Centrolenella colymbiphylum, 63
Centrolenella fleischmanni, 22, 63
Centrolenella valerioi, 63
Ceratophrines, 30
Character evolution, 96–104, 188
Chimpanzee, 26
Chorus: costs and benefits of joining, 172–78; response of to bat predation, 167–68
Christian, K. A., 17
Chuck. *See* Advertisement call, description of
Cladistic analysis. *See* Phylogenetics, to test theories of behavioral evolution
Clutch size, 132–33, 139–41, 156
Cole, L. C., 78, 181
Collias, N. E., 119
Communal nesting, 46–47, 196
Constantine-Paton, M., 21, 112

Constraints: on the evolution of calling behavior, 105–7, 143, 148–150; on the evolution of female choice, 124–26
Conversion of metabolic to acoustic energy: compared among animals 148–50; in *P. pustulosus*, 149–50
Cordosa, A. J., 31
Costs of reproduction, 8, 9, 189; energy, 75–76; 143–62; predation, 77, 163–85. *See also* Energy expended; Predation
Cott, H. B., 163
Cows, dairy, 10
Cricket frog, 21
Crickets, 8, 19, 97
Crinia, 20
Criteria: for selection of a study organism, 25; of female choice, 10–11
Crossley, S., 97, 129
Crump, M. L., 159
Crustaceans, 2
Cutoff frequency, 105

Daniels, R., 14, 75, 146, 182–83
Darwin, C., 1–7, 11, 119, 129, 163
Darwin's frog, 24
Davidson, E. H., 31, 33, 36, 158, 180
Davies, N. B., 3, 11, 16–17, 23, 44, 87, 132, 136–37, 184, 196
Davis, J. W. F., 9
Davison, G. W. H., 3
Dendrobatidae, 24
Derived character, 93; the low-frequency call of *P. pustulosus* as, 102–4. *See also* Apomorphic
Design features of acoustic signals, 119
Dogs, 11
Doherty, 94
Dominance hierarchies, 19
Dominey, W., 9
Downhower, J. F., 93
Drewry, G. E., 72–73, 103, 105, 120
Drosophila melanogaster, 97
Duellman, W. E., 16, 25, 31, 87, 97, 100, 109, 132, 141, 175
Dunbar, R. I. M., 60

Ehrman, L., 10
Eighth cranial nerve, 20. *See also* Auditory system of anurans
Eleutherodactylus, 103
Eleutherodactylus coqui, 11, 93, 124
Eleutherodactylus jasperi, 24

Embleton, T. F. W., 116
Emerson, S. B., 153–55
Emlen, S. T., 14–15, 18, 24, 42, 50, 64, 65, 92, 164, 177, 191
Endler, J. A., 163
Energetic efficiency of calling. *See* Conversion of metabolic to acoustic energy
Energy expended: during calling, 75–76, 144–48, 186, 189; in egg production, 159–61; methods of estimating, 197–201; during noncalling activities, 144–51; physiological limits on the amount of, 151; for reproduction compared to other amphibians and reptiles, 159–61; for reproduction compared among sexes, 156–59, 179–84. *See also* Costs of reproduction, energy
Engystomops, 19, 72
Environmental bioacoustics, 114–20. *See also* Advertisement call, transmission properties of
Esteves, A. E., 31
Eugenic mate choice. *See* Female choice, based on genetic advantages
Evoked vocal responses of males as a behavioral bioassay, 95–96, 109–10, 196
Excess attenuation of acoustic signals, 114–20. *See also* Advertisement call, transmission properties of
Excitatory frequency range of anuran auditory system, 112
Explosive breeders, 14, 44, 64

Factorial scope, 152
Fairchild, L., 16–17
Falconer, P. S., 10, 97
Fellers, G. M., 14, 22, 59, 164
Female choice: in the absence of heritable variation in male traits, 130; accuracy in discriminating among call frequencies, 90–95; in anurans, reviewed, 13–23; based on the advertisement call, 16–23, 55, 84–95, 188; based on the amount of calling, 60; based on avoiding "worst" alleles, 127–28; based on genetic advantages, 18, 123, 126–31, 188–89; based on localizability of the call, 19, 84–87; based on male characters, 2–11; based on male position in the chorus, 59; based on male size, 58; based on natural selection advantages, 7, 9, 126–37, 188; based on non-genetic benefits, 123, 131–37, 189; based

Female choice—*continued*
  on territory quality, 18, 25, 59, 131–35; criteria of, 10–11; as an epiphenomenon of the nervous system, 123–26; movements of female *P. pustulosus* at the breeding site, 41–44, 92–93; neurophysiological constraints on, 138; as a neutrally adaptive trait, 123–26; resulting in conspecific matings, 123
Feng, A. S., 21, 112, 124
Ferner, J. W., 191
Fertilization rates: factors influencing in *P. pustulosus*, 132–37; in anurans, 137; methods of measuring, 196–97
Fibrous masses of vocal cords, 104
Fisher, R. A., 6–11, 19, 104, 128–29, 131, 141–43, 163
Fitness. *See* Male mating success, as a component of fitness; Clutch size
Foam nest: adaptations of, 30, 46–47; construction of, 31; importance in zoogeography, 30; influence on desiccation, 47; influence on predation, 47
Foraging behavior: of *P. pustulosus*, 155 of *T. cirrhosus*, see *T. Cirrhosus*; Predation
Freeman, H. L., 29
Frequency modulation. *See* Advertisement call, frequency modulation of
Frequency sweep. *See* Advertisement call, frequency modulation of
Frequency window of the basilar papilla, 116
Frishkopf, L. S., 16, 21, 138
Frog-eating bat. *See Trachops cirrhosus*
Fuzessery, Z. M., 112

Gallardo, J. M., 31
Gans, C., 107
Gatz, A. J., Jr., 17, 63–64
Gerhardt, H. C., 14, 16, 20–22, 60, 75, 84, 87, 92, 94, 97, 104, 131, 146, 182–83
Genetic correlation of female preference and male trait, 7–10, 123, 142
Genetic covariance. *See* Genetic correlation
Gibson, R. M., 59
Gish, S., 114
Glass frogs, 22
Goldstein, M. H., 138
Gould, S. J., 1
Grass frogs, 112
Green frogs, 96
Greenwalt, C., 71–72
Greer, B. J., 22
Griffin, D. R., 83

Group selection, 5
Growth rates of *P. pustulosus*, 57, 126–28
Guarino, J. L., 15, 49
Guttman, S. I., 13

Haas, R., 163
Hadley, N. F., 144, 160
Halliday, T. R., 1, 16–17, 23, 44, 87, 94, 132, 136–37, 184, 196
Hamilton, W. D., 127, 129, 175
Hamilton, W. J. III., 78
Handicap principle, 9
Harmonics of advertisement call. *See* Advertisement call, harmonic structure of
Hausfater, G., 15
Heffner, H., 77
Heffner, R., 77
Heritability, 7, 9, 19, 143; of the anuran advertisement call, 97; for genetic resistance to pathogens and parasites 129; of sexual display behavior of animals, 97; of male fitness traits, 128–30; 189
Heterospecific clasping in *P. pustulosus*, 42
Heyer, W. R., 28–31, 45–47, 71–73, 99, 105–7, 113, 120, 196
Hillman, S. S., 146
Hinde, R. A., 168
Hogan, J. A., 10, 59
Hogan-Warburg, A. J., 10
Hough, B. R., 31, 33, 36, 158, 180
Howard, R. D., 17–18, 22, 24, 49, 57, 126, 131–32, 142, 163–64, 175, 179–80, 182, 196
Hoy, R. R., 8
Human speech, 104
Hurst, C. J., 23, 93, 96
Huxley, J. S., 5–6, 12
Hyla, 21
Hyla cinerea, 20, 22, 92, 97, 124
Hyla chrysoscelis, 97
Hyla ewingi, 21, 131
Hyla regilla, 18–19
Hyla rosenbergi, 63, 136, 175
Hyla verreauxi, 21, 131
Hyla versicolor, 20, 62
Hylid frogs, 97, 131, 159
Hylidae, 25, 97, 100
Hyoid apparatus, 29

Inferior colliculus, 112
Inhibitory frequency range of the auditory system, 112

Insectivorous bat, 202
Interaural distance, 84
Inverse square law, 114
Irish elk, 1

Jaeger, R. G., 164, 175
Jameson, D. L., 13
Janetos, A. C., 93
Jenni, D. A., 3
Jennson, T. A., 23, 93, 96

Kaplan, R. H., 159
Karr, J. R., 26
Kennedy, J. J., 15, 49
Kirkpatrick, M., 7, 8, 142
Kluge, A. G., 61–62, 68, 132, 136, 175
Knudsen, E. I., 77, 83–84
Konishi, M., 77, 83
Krebs, J. R., 18–19, 22, 59, 164
Kruijt, J. P., 10, 59
Kruse, K. C., 132, 136

Lactate: accumulation during calling and nesting, 147–48; methods of measuring concentration, 200. *See also* Energy expended
Lande, R., 7–8, 10, 109, 129, 142
Larynx, 72
Lawick-Goodall, J., 26–27
Leks, 18
Lenington, S., 10
Leopard frog, 112
Leptodactylidae, 27–29, 99
Leptodactylids, 28, 30, 46, 113
Leptodactylinae, 28–29, 98–103, 121
Leptodacylines, 30, 46
Leptodactylus, 100, 103
Leptodactylus bolivianus, 98
Leptodactylus bufonius, 98
Leptodactylus chaquensis, 98
Leptodactylus fuscus, 99
Leptodactylus gracilis, 99
Leptodactylus labialus, 99
Leptodactylus latinasus, 99
Leptodactylus melanonotus, 99
Leptodactylus mystaceus, 99
Leptodactylus ocellatus, 99
Leptodactylus pentadactylus, 99–101, 114, 172–73
Leptodactylus podicipinus, 99
Leptodactylus poecilochilus, 99
Leptodactylus syphax, 99
Leptodactylus wagneri, 99

Leptodiera annulata, 175
Lewis, B., 78
Lewontin, R. C., 9, 143, 182
Licht, L. E., 16, 132
Licht, P., 151, 153–54
Life-history patterns, 180–82
Linkage disequilibrium, 8. *See also* Genetic correlation
Liopelmatid, 30
Littlejohn, M. J., 16, 20–21, 131, 194
Live-bearing frog, 24
Loftus-Hills, J. J., 16, 20–21
Long-tailed widow bird, 130
Lorenz, K. L., 97, 168
Lutz, B., 24
Lynch, J. D., 28–30, 46, 99, 108–9

McClanahan, L. L., 146
McDiarmid, R. W., 63
MacDonald, J., 97, 129
MacNally, R. C., 149, 159–60
Maier, V., 119
Male competition, 3–4, 17; responses to vocalizations of other males, 73–75, 95–96
Male contribution to female reproductive success. *See* Fertilization rates
Male mating success: coefficients of variation in, 61; compared to that of other anurans, 62–64; as a component of fitness, 15, 189; methods of measuring, 191–92; in *P. pustulosus*, 49–65
Male size: and age, *see* Growth rates; influence on the advertisement call, *see* Advertisement call, differences based on male size; influence on mating success, *see* Mating success
Mate choice. *See* Female choice
Mating call, 19; *see also* Advertisement call
Mating call detector, 113
Marler, P., 77–78, 83, 114, 117, 119, 163
Marten, K., 114, 117
Martin, A. A., 46
Martin, W. F., 16, 72, 87, 97, 100, 104–5, 107
Martof, B. S., 24
Maximum aerobic capacity, 153
Maynard Smith, J., 6, 8–9, 11, 18
Mayr, E., 1, 3–4, 10–11, 75, 97, 143
Mecham, J. S., 16, 20–21
Mechanical cost of calling, 147
Metabolic performance, 154
Metabolic scope, 152–53
"Mew" calls, 37

Michaud, T. C., 20, 194
Michelson, A., 60, 116–17
Micronycteris megalotis, 169–70, 202–3
Milstead, W. W., 31
Mobbing calls, structure of, 77–78
Moffat, A. J. M., 21–22, 93, 109, 112–13, 124
Moonlight, influence of on predator detection, 166–67
Morton, E. S., 114, 117, 119, 150
Motivation structural rules, 119
Mounce, M., 132, 136
Moynihan, M., 77, 163
Mudry, K. M., 21, 112–13

Nagy, K. A., 160–61
Narins, P. M., 16, 21, 93, 111–12, 124
Natterjack toad, 23
Natural history, summary of previous studies of *P. pustulosus*, 30–31
Natural selection, 123, 188; as an analogy of artificial selection, 4; contrasted with sexual selection, 2, 7
Nest building: duration of, 46; energy expended during, 150–51; lactate accumulation during, 150–51
Nest site selection, 46
Nevo, E., 16, 21
Nice, M. M., 48
Noble, G. K., 5–6, 12, 28, 42
Nonbreeding mortality, 181–82
Noncalling behaviors, energy expended during, 146–47, 183–84. *See also* Alternative male reproductive behaviors

O'Donald, P., 1, 5, 8–9, 128
Oldham, R. S., 16
Olsen, N., 116
Oogenesis, patterns of in *P. pustulosus*, 31–34, 36
Operational sex ratio, 51–54, 64
Orians, G. H., 10
Oring, L. W., 14, 50, 64–65, 177
O'Rourke, F. J., 1
Ortleb, E. P., 19, 31, 46
Otte, D., 1, 7
Outgroup comparison, 102–4; 188
Oxygen consumption: methods of measuring, 197–201; rate of during calling, 76–77, 144–48; rate of during noncalling activities, 144–51. *See also* Energy expended

Parasitic mating strategies. *See* Alternative male reproductive behaviors
Parker, G. A., 4, 44, 130, 180
Partridge, L., 130
Passmore, N. I., 60
Paternal certainty, 15, 49
Paterson, H. E. H., 131
Paul, R. C., 8
Payne, K., 63–64
Payne, R. R., 63–64
Peacocks, 1
Peripheral auditory system, 20. *See also* Anuran auditory system
Peripheral males, 14. *See also* Alternative male reproductive behaviors
Perrill, S. A., 14, 75, 146, 182–83
Philander opossum, 172–73
Phyllostomatidae, 201
Phylogenetics, to test theories of behavioral evolution, 96–104, 188
Physalaemus aguirrei, 108–9
Physalaemus albonatus, 30, 98, 109
Physalaemus barbouri, 30
Physalaemus bilogonigerus, 30–31, 98, 109
Physalaemus centralis, 109
Physalaemus cicada, 30
Physalaemus cuvieri, 30–31, 98, 108–9
Physalaemus fernandezae, 30–31, 98, 109
Physalaemus fuscomaculatus, 31, 109
Physalaemus gracilis, 30–31, 98, 105, 107, 109
Physalaemus henseli, 30–31, 98, 109
Physalaemus jordensis, 30
Physalaemus maculiventris, 98, 109
Physalaemus obtectus, 109
Physalaemus olfersi, 98, 107
Physalaemus petersi, 31
Physalaemus riograndensis, 30–31, 98, 109
Physalaemus santafecinus, 30–31, 98, 108–9
Physalaemus signiferus, 98, 108–9
Pianka, E. R., 180
Piercy, J. E., 116
Pinnipeds, 179
Plesiomorphic, 102
Pleszczynska, 10, 88
Pleurodema, 28
Potamocarcinus richmondi, 172
Pough, F. H., 153–55, 160
Poulton, E. B., 3
Predation: influence of sound localization on, 78–79; influence of signal structure on,

119–20; on males during courtship, 77, 163–85; by *Trachops cirrhosus* on *P. pustulosus*, 77–78, 163–85; by use of acoustic cues, 77–78
Predator detection, 166–71; methods of measuring, 202–3
Preferential phonotaxis by females, methods of determining, 194–95. *See also* Female choice, based on differences in advertisement call
Pressure gradient receptor, 77
Pressure receptor, 77
Prey items, found in the stomachs of *P. pustulosus*, 155
Prolonged breeders, 14, 64
Pseudacris, 20
Pseudemys scripta, 175
Púngala frog, 36
Pyles, R. A., 16, 87, 97, 100

Quine, D. B., 114, 117

Radiators, sound, efficiency of, 106–7
Ramer, J. D., 23, 93, 96
Ranas, 27
Rana catesbeiana, 63
Rana clamitans, 18, 62, 83
Rana sylvatica, 17, 62, 136
Rand, A. S., 27, 31, 33, 35, 37, 45–46, 66, 71–79, 83–84, 88, 103, 105, 107, 110, 120, 144, 148–49, 158, 160, 196
Rand, W. M., 33, 144
Ranidella parinsignifera, 159–160
Ranidella signifera, 159–60
Red-eyed treefrog, 25, 27, 31. *See also* *Agalychnis calidryas*
Red-winged blackbird, 15, 115
Rejection mode, of auditory system, 112
Release calls, 41
Reproductive character displacement, 131
Reproductive success: as a component of fitness, 49; defined, 49; estimations of, 49. *See also* Mating success
Resler, C., 84
Rheinlaender, J., 84
Rhinoderma darwini, 24. *See also* Darwin's frog
Richards, D. G., 60, 113–15, 117, 119
Rivero, J. A., 31
Roeder, K. D., 169
Roederer, J. G., 75, 87, 104
Rohlf, F. J., 91, 192, 202

Rossing, T. D., 39, 104–5
Roth, H. D., 10
Rubenstein, D. I., 16–17
Rufous-sided towhee, 115
Runaway sexual selection. *See* Sexual selection, runaway
Ruse, M., 3–5
Ryan, M. J., 24, 37–38, 53, 56, 68, 73, 79, 83–84, 89–90, 110, 114, 133–35, 137, 145–50, 153, 158, 160, 166–67, 169, 171, 173–74, 199

Salthe, S. N., 16, 132, 141
Satellite males. *See* Alternative male reproductive behaviors
Savage, J. M., 30
Scaphiopus hammondi, 153
Schlegel, P., 112
Schoener, A., 163
Schoener, T. W., 163
Schwartz, J. J., 22
Scramble competition, 13, 17, 186
Searcy, W. A., 10, 18, 129, 131
Secondary sexual characteristics, 2
Selander, R. K., 2, 10, 163
Selection, contrasted with evolutionary response to selection, 9, 97
Selfish herd theory, 176
Self-reinforcing choice, 6
Servicing time, 65, 144
Sexton, O. J., 19, 31, 46
Sexual dimorphism, 1–2, 179–84
Sexual selection: criticisms of, 4–6, 9–10; effect on call evolution, 96–104; estimating the intensity of, 61–63; evolution of maladaptive traits through, 8; Fisher's contribution to the theory of, 6–10; genetic assumptions of, 9; genetic models of, 7; opposed by the selective force of predation, 163–85; review of Darwin's theory of, 2–6; runaway, 6–10, 123, 187–89
Seymour, R. S., 146, 193
Shalter, M. D., 78
Shields, W. M., 22
Shine, R., 179
Shoemaker, V. H., 146
Siegel, S., 192, 196, 202–4
Smith, G. C., 159
Smith, W. J., 77, 163
Smith-Gill, S. J., 132, 136
Snyder, G. K., 145
Sokal, R. R., 91, 192, 195, 202

Sound localization: accuracy of in *P. pustulosus*, 84; by anurans, 84; influenced by call complexity, 84; influenced by frequency modulation of the call, 113; influence on predation, 77–78
Sound pressure level. See Advertisement call, intensity
Sound radiator, 105, 150
Sound resonator, 105
South American bullfrog, 114, 172, 175. See also *Leptodactylus pentadactylus*
Spadefoot toads, 21, 153
Species-isolating mechanisms, 21
Spherical spreading of sound, 114
Statistical analyses, 192, 195–97
Stearns, S. C., 143, 180, 182
Steele, R. G. D., 135, 192, 196–97, 201
Stinner, J. N., 146
Straughn, I. R., 71, 99, 113
Suga, N., 112
Sullivan, B. K., 17, 19, 64, 93, 184

Taft, L. K., 82–83, 165–69, 171–74
Taigen, T. L., 193–95
Tannenbaum, B. R., 15
Taxonomy, of *P. pustulosus*, 27–30
Taylor, D. H., 13
Teleology, as an approach to studies of behavioral evolution, 143
Telmatobiinae, 100, 103
Territory, defense of in *P. pustulosus*, 48
Thornhill, R., 1, 11, 131
Tilley, S. G., 144
Tinbergen, N., 167–68
Tinkle, D. W., 144, 150
Toe-clipping, 191
Toft, C. A., 155
Torrie, J. H., 135, 192, 196–97, 201
Torus semicircularis, 112
Townsend, D. S., 49
Trachops cirrhosus: preference of for complex advertisement calls, 78–84; experimental methods to determine preference of for complex advertisement calls, 201–2; predation by on *P. pustulosus*, 78–83, 120, 164–73, 178, 201–3
Tracy, C. R., 17
Transmission distance of components of the advertisement call, 114–15
Treefrogs, 21, 26, 72, 93
Trivers, R. L., 8, 10, 126–27, 143–44, 163
Tuttle, M. D., 78–79, 83, 164–69, 171–74
Two-tone suppression, 112–13

Uta stansburiana, 160–61

Vanzolinius discodactylus, 99
Verhencamp, S. L., 172
Vial, J. L., 13
Villa, J., 31, 36
Vocal cords: fibrous masses of, 72; vibrations of, 72. See also Advertisement call, production of
Vocal sacs: correlation with type of calling site, 107; influence of radiation of sound, 106–7

Wade, M. J., 9, 14, 61, 94
Waist bands for individual identification during behavioral observations, 191
Waldman, B., 17, 184
Walkowiak, W., 21, 112
Wallace, A. R., 4–6, 12
Waser, M. S., 114–15, 119
Waser, P. M., 114–15, 119
Weathers, W. W., 145
Weismann, A., 4, 11
Wells, K. D., 13, 14, 16, 18–19, 22, 24, 44, 63–64, 172, 177
West Eberhard, M. J., 1, 7, 9, 41, 96, 109, 129–30, 142
Whine. See Advertisement call, description of
White-lined bat, 15
Whitney, C. L., 18–19, 22, 59, 164
Wilbur, H. M., 16–17, 144
Wilczynski, W., 84
Wiley, E. O., and R. H., 60, 102, 113–15, 117, 172, 189
Williams, G. C., 5, 126, 129, 143
Wilson, E. O., 48, 143
Withers, P. C., 146
Wittenberger, J. F., 177
Wood frogs, 182
Woolbright, L. L., 179
Wrangham, R. W., 11
Wrestling bouts among male *P. pustulosus*, 41

Yarnell, R. M., 19, 31
Young, D., 149

Zahavi, A., 8, 9, 128
Zelick, R. D., 93
Zoogeography of *P. pustulosus*, 27–30
Zuk, M. Z., 127, 129
Zweifel, R. G., 18, 87